SELECTED SOLUTIONS MANUAL

Matthew E. Johll

Illinois Valley Community College

INTRODUCTORY CHEMISTRY

FOURTH EDITION

Nivaldo J. Tro

Prentice Hall

Boston Columbus Indianapolis New York San Francisco Upper Saddle River
Amsterdam Cape Town Dubai London Madrid Milan Munich Paris Montréal Toronto
Delhi Mexico City São Paulo Sydney Hong Kong Seoul Singapore Taipei Tokyo

Editor in Chief, Chemistry: Adam Jaworski
Marketing Manager: Erin Gardner
Senior Project Editor: Jennifer Hart
Assistant Editor: Lisa Pierce
Managing Editor, Chemistry and Geosciences: Gina M. Cheselka
Project Manager, Science: Shari Toron
Senior Manufacturing & Operations Manager: Nick Sklitsis
Operations Specialist: Maura Zaldivar
Supplement Cover Designer: Paul Gourhan
Cover Illustration: Quade Paul

Pearson Prentice Hall
Upper Saddle River, NJ 07458

Printed in the United States of America

10 9 8 7 6 5 4 3 2 1

ISBN-13: 978-0-321-73018-3
ISBN-10: 0-321-73018-6

Prentice Hall
is an imprint of

www.pearsonhighered.com

Student Solutions Manual

Table of Contents

The Chemical World

1

Questions

1. Soda contains carbon dioxide molecules that are forced into a mixture with water molecules because of the increased pressure. Opening the can of soda will release the pressure, which allows the carbon dioxide, to escape from the mixture and form bubbles.

3. Chemists study the world around us and try to explain how it works. By studying the interactions of atoms and molecules, chemists hope to better understand the world around us.

5. Chemistry: The science that seeks to understand what matter does by studying what atoms and molecules do.

7. The scientific method is a way of approaching a problem that emphasizes making observations, planning experiments, performing experiments, and then using logic to come to a conclusion.

9. A scientific law is a brief statement that summarizes previous observations and can be used to predict the results of future related experiments. The scientific law does not explain why the observations occur. That is the role of a scientific theory.

11. A scientific theory is the best current explanation of a phenomenon that has been validated by years' worth of experiments.

13. John Dalton proposed his atomic theory, which states that all matter is composed of small, indestructible particles called atoms.

Problems

15. Carbon dioxide contains one carbon atom and two oxygen atoms. Water contains one oxygen atom and two hydrogen atoms.

17. a) observation
    b) theory
    c) law
    d) observation

19. Examination of the data table reveals that when the mass of the gas increases, a corresponding increase in the volume of the gas occurs. Two equivalent ways of expressing this information as a scientific law would be: (1) the mass of a gas is directly proportional to its volume or (2) the mass to volume ratio of a gas is constant.

21. a) The reactivity of a chemical is proportional to the molecular size.
    b) Reactivity is due to the momentum of collisions, and because the larger molecules have a greater momentum, they have a greater reactivity.
    Remember this problem is for an imaginary universe!

Measurement and Problem Solving

2

Questions

1. Units must be included with a number so that there is no doubt as to the meaning of the number. For example, if you reported a mass of 5, that would be unclear. However, if you reported a mass of 5 grams, there would be no confusion.

3. Scientific notation is used to make very small numbers and very large numbers easier to write and to understand.

5. a) Zeros in the middle of two nonzero numbers ARE significant.
   b) Zeros at the end of a number with a decimal point ARE significant.
   c) Zeros at the beginning of a number are NOT significant.
   d) Zeros at the end of a number with no decimal point are ambiguous. Avoid using these numbers and use scientific notation instead.

7. The number of significant digits in multiplication and division problems is determined by whichever number contains the fewest significant digits.

9. When calculations involve a combination of multiplication, division, addition, and subtraction, it is necessary to do the mathematical steps inside the parentheses first. Then, determine how many significant digits there should be during each step, but instead of rounding, underline the last significant digit. Keep all digits in the problem until it is completely finished, then round based on the significant figure rules.

11. The SI unit for length is the meter (m), for mass is the kilogram (kg), and for time is the second (s).

13. The frisbee would be measured in meters with a prefix of deci (0.1). One might choose to use a prefix of centi (0.01), as it tends to be more commonly used.

15. Correct answers may vary based on the marking division of the ruler.

17. Units can act as a guide in the calculation and are able to show if the calculation is off track. More importantly, the units provide important information about the final answer and must be included if the answer is to be correctly written and understood.

19. A conversion factor is a fraction composed of two equivalent quantities, and is used to convert information from one set of units to another.

21. inches → feet   feet → inches

$\frac{1\text{ foot}}{12\text{ inches}}$ $\frac{12\text{ inches}}{1\text{ foot}}$

23. Sort: Sort the information into the *given* information and what the problem is asking you to *find.*
Strategize: You formulate a series of steps that form a solution map.
Solve: Carry out the mathematical operations set up by the solution map.
Check: Verify that the answer makes sense and has the correct units.

25. grams → pounds

$\frac{1\text{ lb}}{453.6\text{ g}}$

27. meters → centimeters → inches → feet

$\frac{100\text{ cm}}{1\text{ m}}$ $\frac{1\text{ in}}{2.54\text{ cm}}$ $\frac{1\text{ ft}}{12\text{ in}}$

29. Density is the ratio of the mass of an object to the volume of the object. Objects that have a large density are perceived to be heavy. Objects that have a small density are perceived to be light. Density can work as a conversion factor because it is an equality that converts between mass and volume. For example, 1 gram of water will occupy a volume of 1 mL. Therefore if you have a mass of 100 grams you can use the density to calculate the volume.

Problems

Scientific Notation

31. a) 3.6458×10^{7}
b) 1.286×10^{6}
c) 1.949×10^{7}
d) 5.32×10^{5}

33. a) 7.461×10^{-11} m
b) 1.58×10^{-5} mi
c) 6.32×10^{-7} m
d) 1.5×10^{-5} m

35. a) 602,200,000,000,000,000,000,000 atoms
b) 0.00000000000000000016 C
c) 299,000,000 m/s
d) 344 m/s

37. a) 32,200,000
    b) 0.0072
    c) 118,000,000,000
    d) 0.00000943

39.

2,000,000,000	2×10^{9}
1,211,000,000	1.211×10^{9}
0.000874	8.74×10^{-4}
320,000,000,000	3.2×10^{11}

Significant Figures

41. a) 54.9 mL
    b) 48.7 °C
    c) 46.83 °C
    d) 64 mL

43. a) 0.005050 m
    b) 0.00000000000000060 s
    c) 220,103 kg
    d) 0.00108 in.

45. a) 4
    b) 4
    c) 6
    d) 5

47.

Number	Significant Figures	
a) 895675	6	*correct*
b) 0.000869	6	*incorrect, 3: leading zeros are not significant*
c) 0.5672100	5	*incorrect, 7: terminal zeros are significant*
d) $6.022x10^{23}$	4	*correct*

Rounding

49. a) 256.0
    b) 0.0004893
    c) 2.901×10^{-4}
    d) 2.231×10^{6}

51. a) 2.3
    b) 2.4
    c) 2.3
    d) 2.4

53. a) incorrect, 42.3
    b) correct
    c) correct
    d) incorrect, 0.0456

55.

Number	Rounded to 4 significant figures	Rounded to 2 significant figures	Rounded to 1 significant figure
1.45815	1.458	1.5	1
8.32466	8.325	8.3	8
84.57225	84.57	85	8×10^1
132.5512	132.6	1.3×10^2	1×10^2

Significant Figures in Calculations

57. a) 0.054
    b) 0.619
    c) 1.2×10^8
    d) 6.6

59. a) incorrect, 4.22×10^3
    b) correct
    c) incorrect, 3.9969
    d) correct

61. a) 110.6
    b) 41.4
    c) 183.3
    d) 1.22

63. a) correct
    b) incorrect, 1.0982
    c) correct
    d) incorrect, 3.53

65. a) 3.9×10^3
    b) 632
    c) 8.93×10^4
    d) 6.34

67. a) incorrect, 3.15×10^3
    b) correct
    c) correct
    d) correct

Unit Conversions

69. a) $3.55 \text{ kg}\times\frac{1000 \text{ g}}{1 \text{ kg}} = 3.55\text{x}10^3 \text{ g}$

    b) $8944 \text{ mm}\times\frac{1 \text{ m}}{1000 \text{ mm}} = 8.944 \text{ m}$

    c) $4598 \text{ mg}\times\frac{1 \text{ g}}{1000 \text{ mg}}\times\frac{1 \text{ kg}}{1000 \text{ g}} = 4.598\times10^{-3}\text{kg}$

    d) $0.0187 \text{ L}\times\frac{1000 \text{ mL}}{1 \text{ L}} = 18.7 \text{ mL}$

71. a) $5.88 \text{ dL}\times\frac{1 \text{ L}}{10 \text{ dL}} = 0.588 \text{ L}$

    b) $3.41\times10^{-5} \text{ g}\times\frac{1{,}000{,}000 \text{ µg}}{1 \text{ g}} = 34.1 \text{ µg}$

    c) $1.01\times10^{-8} \text{ s}\times\frac{1\text{x}10^9 \text{ ns}}{1 \text{ s}} = 10.1 \text{ ns}$

    d) $2.19 \text{ pm}\times\frac{1 \text{ m}}{1\text{x}10^{12}\text{pm}} = 2.19\times10^{-12} \text{ m}$

73. a) $22.5 \text{ in}\times\frac{2.54 \text{ cm}}{1 \text{ in}} = 57.2 \text{ cm}$

    b) $126 \text{ ft}\times\frac{12 \text{ in}}{1 \text{ ft}}\times\frac{2.54 \text{ cm}}{1 \text{ in}}\times\frac{1 \text{ m}}{100 \text{ cm}} = 38.4 \text{ m}$

    c) $825 \text{ yd}\times\frac{1 \text{ m}}{1.094 \text{ yd}}\times\frac{1 \text{ km}}{1000 \text{ m}} = 0.754 \text{ km}$

    d) $2.4 \text{ in}\times\frac{2.54 \text{ cm}}{1 \text{ in}}\times\frac{10 \text{ mm}}{1 \text{ cm}} = 61 \text{ mm}$

75. a) $40.0\text{ cm}\times\frac{1\text{ in}}{2.54\text{ cm}}=15.7\text{ in}$

b) $27.8\text{ m}\times\frac{39.37\text{ in}}{1\text{ m}}\times\frac{1\text{ ft}}{12\text{ in}}=91.2\text{ ft}$

c) $10.0\text{ km}\times\frac{0.6214\text{ mi}}{1\text{ km}}=6.21\text{ mi}$

d) $3845\text{ kg}\times\frac{2.205\text{ lb}}{1\text{ kg}}=8478\text{ lb}$

77.

m	km	Mm	Gm	Tm
5.08×10^{8} m	5.08×10^{5} km	508 Mm	0.508 Gm	5.08×10^{-4} Tm
2.7976×10^{10} m	2.7976×10^{7} km	27,976 Mm	27.976 Gm	2.7976×10^{-1}Tm
1.77×10^{12} m	1.77×10^{9} km	1.77×10^{6} Mm	1.77×10^{3} Gm	1.77 Tm
1.5×10^{8} m	1.5×10^{5} km	1.5×10^{2} Mm	0.15 Gm	1.5×10^{-4} Tm
4.23×10^{11} m	4.23×10^{8} km	4.23×10^{5} Mm	423 Gm	0.423 Tm

79. a) $2.255\times10^{10}\text{g}\times\frac{1\text{ kg}}{1000\text{ g}}=2.255\times10^{7}\text{ kg}$

b) $2.255\times10^{10}\text{g}\times\frac{1\text{ Mg}}{1\times10^{6}\text{ g}}=2.255\times10^{4}\text{ Mg}$

c) $2.255\times10^{10}\text{g}\times\frac{1000\text{ mg}}{1\text{ g}}=2.255\times10^{13}\text{ mg}$

d) $2.255\times10^{10}\text{g}\times\frac{1\text{ kg}}{1000\text{ g}}\times\frac{1\text{ metric ton}}{1000\text{ kg}}=2.255\times10^{4}\text{ metric tons}$

81. $3.3\text{ lb}\times\frac{1\text{ kg}}{2.205\text{ lb}}\times\frac{1000\text{g}}{1\text{ kg}}=1.5\times10^{3}\text{g}$

83. $10.0\text{ km}\times\frac{0.6214\text{ mi}}{1\text{ km}}\times\frac{1\text{ hr}}{7.5\text{ mi}}\times\frac{60\text{ min}}{1\text{ hr}}=5.0\times10^{1}\text{ min}$

85. $5.0\text{ qt}\times\frac{1\text{ L}}{1.057\text{ qt}}\times\frac{1000\text{ cm}^{3}}{1\text{ L}}=4.7\times10^{3}\text{cm}^{3}$

Units Raised to a Power

87. a) $1.0\ \text{km}^2 \times \frac{(1000\ \text{m})^2}{(1\ \text{km})^2} = 1.0 \times 10^6\ \text{m}^2$

b) $1.0\ \text{cm}^3 \times \frac{(1\ \text{m})^3}{(100\ \text{cm})^3} = 1.0 \times 10^{-6}\ \text{m}^3$

c) $1.0\ \text{mm}^3 \times \frac{(1\text{m})^3}{(1000\ \text{mm})^3} = 1.0 \times 10^{-9}\ \text{m}^3$

89. a) $6.2 \times 10^{-31}\ \text{m}^3 \times \frac{(1\ \text{pm})^3}{(1 \times 10^{-12}\text{m})^3} = 6.2 \times 10^5 \text{pm}^3$

b) $6.2 \times 10^{-31}\ \text{m}^3 \times \frac{(1\ \text{nm})^3}{(1 \times 10^{-9}\text{m})^3} = 6.2 \times 10^{-4} \text{nm}^3$

c) $6.2 \times 10^{-31}\ \text{m}^3 \times \frac{(1\text{Å})^3}{(1 \times 10^{-10}\text{m})^3} = 6.2 \times 10^{-1}\ \text{Å}^3$

91. a) $215\ \text{m}^2 \times \frac{(1\ \text{km})^2}{(1000\ \text{m})^2} = 2.15 \times 10^{-4}\ \text{km}^2$

b) $215\ \text{m}^2 \times \frac{(10\ \text{dm})^2}{(1\ \text{m})^2} = 2.15 \times 10^4\ \text{dm}^2$

c) $215\ \text{m}^2 \times \frac{(100\ \text{cm})^2}{(1\ \text{m})^2} = 2.15 \times 10^6\ \text{cm}^2$

93. 954 million acres = 954,000,000 acres = 9.54×10^8 acres;

$9.54 \times 10^8\ \text{acres} \times \frac{43{,}560\ \text{ft}^2}{1\ \text{acre}} \times \frac{(1\ \text{mi})^2}{(5280\ \text{ft})^2} = 1.49 \times 10^6\ \text{mi}^2$

Density

95. $d = \frac{35.4\ \text{g}}{3.11\ \text{cm}^3} = 11.4\ \text{g/cm}^3$; This matches the density of lead.

97. $d = \frac{3.15 \times 10^3\ \text{g}}{2.50\ \text{L}} \times \frac{1\ \text{L}}{1000\ \text{mL}} \times \frac{1\ \text{mL}}{1\ \text{cm}^3} = 1.26\ \text{g/cm}^3$

99. $d = \frac{206 \text{ grams}}{10.7 \text{ mL}} \times \frac{1 \text{ mL}}{1 \text{ cm}^3} = 19.3 \text{ g/cm}^3$;

Yes, the density matches gold.

101. a) $387 \text{ mL} \times \frac{1 \text{ cm}^3}{1 \text{ mL}} \times \frac{1.11 \text{g}}{1 \text{ cm}^3} = 4.30 \times 10^2 \text{g}$

b) $3.46 \text{ kg} \times \frac{1000 \text{ g}}{1 \text{ kg}} \times \frac{1 \text{ cm}^3}{1.11 \text{g}} \times \frac{1 \text{ mL}}{1 \text{ cm}^3} \times \frac{1 \text{ L}}{1000 \text{ mL}} = 3.12 \text{ L}$

Cumulative Problems

103. a) $d = \frac{m}{V}$ and $m = d \times V$

$m_{gold} = 19.3 \frac{\text{g}}{\text{cm}^3} \times 1.75 \text{ L} \times \frac{1000 \text{ cm}^3}{1 \text{ L}} = 3.38 \times 10^4 \text{g}$

$m_{sand} = 3.00 \frac{\text{g}}{\text{cm}^3} \times 1.75 \text{ L} \times \frac{1000 \text{ cm}^3}{1 \text{ L}} = 5.25 \times 10^3 \text{g}$

b) Yes, the thief set off the alarm because the sand was much lighter than the gold vase.

105. $d = \frac{m}{V} = \frac{5.14 \text{lb.}}{13.4 \text{in.}^3} \times \frac{453.6 \text{g}}{1 \text{lb.}} \times \left(\frac{1 \text{ in.}}{2.54 \text{ cm}}\right)^3 = 10.6 \text{ g/cm}^3$

107. $d = 2.7 \frac{\text{g}}{\text{cm}^3} \times \frac{1 \text{ kg}}{1000 \text{ g}} \times \frac{(100 \text{ cm})^3}{(1 \text{ m})^3} = 2.7 \times 10^3 \text{ kg/cm}^3$

109. $150 \text{ yd}^3 \times \frac{(1 \text{ m})^3}{(1.094 \text{ yd})^3} \times \frac{(100 \text{ cm})^3}{(1 \text{ m})^3} \times \frac{1.0 \text{ g}}{1 \text{ cm}^3} \times \frac{1 \text{ lb}}{453.6 \text{ g}} = 2.5 \times 10^5 \text{ lbs}$

111. $155{,}211 \text{ L} \times \frac{1000 \text{ cm}^3}{1 \text{ L}} \times \frac{0.768 \text{ g}}{1 \text{ cm}^3} \times \frac{1 \text{ kg}}{1000 \text{g}} = 1.19 \times 10^5 \text{ kg}$

113. $\frac{43 \text{ mi}}{1 \text{ gal}} \times \frac{1 \text{ gal}}{3.785 \text{ L}} \times \frac{1 \text{ km}}{0.6214 \text{ mi}} = 18 \text{ km/L}$

115. $76.5 \text{ L} \times \frac{1 \text{ gal}}{3.785 \text{ L}} \times \frac{38 \text{ mi}}{1 \text{ gal}} = 7.7 \times 10^2 \text{ mi}$

117. Because the block A has the larger mass and the lesser volume, and the density is the ratio of mass divided by volume, the density of block A will be greater than the density of block B.

119. Mass of cylinder 1 = 1.35 × Mass cylinder 2

Volume of cylinder 1 = 0.792 × Volume cylinder 2

Density of cylinder 1 = 3.85 g/cm^3

Density of cylinder 2 = ?

$$D_1 = \frac{M_1}{V_1} = \frac{1.35 \times M_2}{0.792 \times V_2} = 3.85 \text{ g/cm}^3$$

$$\frac{M_2}{V_2} = \frac{0.792}{1.35} \times 3.85 \text{g/cm}^3 = 2.26 \text{ g/cm}^3$$

Highlight Problems

121. $1.55 \times 10^5 \text{ft} \times \frac{0.3048 \text{ m}}{1 \text{ ft}} \times \frac{1 \text{ km}}{1000 \text{ m}} = 47.2 \text{ km}$

$155 \text{ km} - 47.2 = 108$ km difference in altitude.

The orbiter would have attempted to establish an orbit 47.2 km above Mars.

123. $m = (1 \times 10^3 \text{suns})(2.0 \times 10^{30} \text{kg/sun}) \frac{1000 \text{ g}}{1 \text{ kg}} = 2.0 \times 10^{36} \text{g}$

$$V = 4/3(3.14)\left(\frac{2.16 \times 10^3 \text{mi}}{2}\right)^3 = 5.27 \times 10^9 \text{mi}^3$$

$$V = 5.27 \times 10^9 \text{mi}^3 \times \frac{(1 \text{ km})^3}{(0.6214 \text{ mi})^3} \times \frac{(1000 \text{ m})^3}{(1 \text{ km})^3} \times \frac{(100 \text{ cm})^3}{(1 \text{ m})^3} = 2.20 \times 10^{25} \text{cm}^3$$

$$d = \frac{2.0 \times 10^{36} \text{g}}{2.20 \times 10^{25} \text{cm}^3} = 9.1 \times 10^{10} \text{ g/cm}^3$$

Matter and Energy

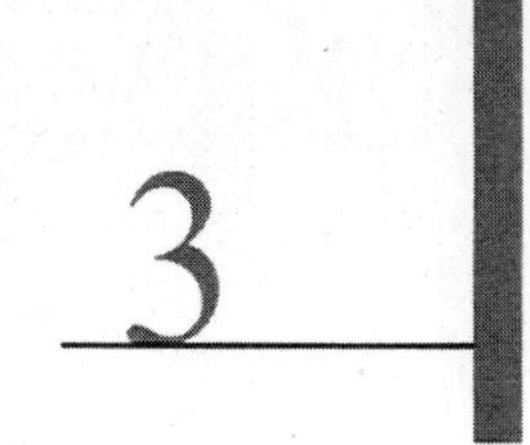

Questions

1. Matter is defined as anything that occupies space and has mass. It can be thought of as the physical material that makes up the universe.

3. Matter can be found as either a solid, a liquid, or a gas.

5. A crystalline solid has atoms arranged in a repeating geometric pattern, whereas an amorphous solid has atoms that do not form repeating patterns.

7. A gas has no fixed volume or shape. Rather, it assumes both the shape and the volume of the container it occupies.

9. A mixture is composed of two or more pure substances that have been mixed together in a variable proportion.

11. A pure substance is composed of only one type of atom or one type of molecule.

13. A mixture is formed when two or more pure substances are mixed together, but a new substance is not formed. A compound is formed when two or more elements are bonded together to form a new substance.

15. A physical change is one that does not alter the chemical composition of a compound. A chemical change, however, will alter the chemical composition of a compound.

17. Energy is the capacity to do work or generate heat.

19. Kinetic energy is the energy associated with motion. Potential energy is the energy associated with position or composition.

21. The common unit for energy in the laboratory is the joule (J). Another commonly used unit of energy found in nutritional information is the calorie (cal) and the Calorie (Cal). Finally, another common unit of energy is the kilowatt-hour (kWh) which is how electricity is measured from utilities.

23. An endothermic reaction is a chemical reaction that absorbs energy. The products have greater energy than the reactants.

25. Temperature is the measure of thermal energy of matter whereas heat is the transfer or exchange of thermal energy caused by a temperature difference.

27. Heat capacity is a measure of how much heat is needed to raise the temperature of a given substance by 1 °C.

29. $°C = \frac{[°F - 32]}{1.8} \Rightarrow 1.8 \times °C = [°F - 32] \Rightarrow °F = (1.8 \times °C) + 32$

Problems

Classifying Matter

31. a) element
    b) element
    c) compound
    d) compound

33. a) homogeneous
    b) heterogeneous
    c) homogeneous
    d) homogeneous

35. a) pure, element
    b) mixture, homogeneous
    c) mixture, heterogeneous
    d) mixture, heterogeneous

Physical and Chemical Properties and Physical and Chemical Changes

37. a) chemical
    b) physical
    c) physical
    d) chemical

39. Physical Properties: colorless, odorless, gas at room temperature, one liter has a mass of 1.260 g under standard conditions, mixes with acetone
    Chemical Properties: flammable, polymerizes to form polyethylene

41. a) chemical
    b) physical
    c) chemical
    d) chemical

43. a) physical
    b) chemical

The Conservation of Mass

45. Mass of reactants = Mass of products; 42 kg + 168 kg = 210 kg

47. a) Yes. 67.5 g reactants = 67.5 g products
    b) No. 303.5 g reactants ≠ 294 g products

49. Mass of reactants = Mass of products
    9.7 g + 34.7g = 29.3g + g water →
    g water = 44.4 - 29.3 = 15.1 g

Conversion of Energy Units

51. a) calories → joules

$$588 \text{ cal} \times \frac{4.18 \text{ J}}{1 \text{ cal}} = 2.46\text{x}10^3 \text{ J}$$

b) joules → calories → Calories

$$17.4 \text{ J} \times \frac{1 \text{ cal}}{4.18 \text{ J}} \times \frac{1 \text{ Cal}}{1000 \text{ cal}} = 4.16\text{x}10^{-3} \text{Cal}$$

c) kilojoules → joules → calories → Calories

$$134 \text{ kJ} \times \frac{1000 \text{ J}}{1 \text{ kJ}} \times \frac{1 \text{ cal}}{4.18 \text{ J}} \times \frac{1 \text{ Cal}}{1000 \text{ cal}} = 32.1 \text{ Cal}$$

d) Calories → calories → joules

$$56.2 \text{ Cal} \times \frac{1000 \text{ cal}}{1 \text{ Cal}} \times \frac{4.18 \text{ J}}{1 \text{ cal}} = 2.35\text{x}10^5 \text{ J}$$

53. a) $25 \text{ kWh} \times \frac{3.60\text{x}10^6 \text{ J}}{1 \text{ kWh}} = 9.0\text{x}10^7 \text{ J}$

b) $249 \text{ cal} \times \frac{1 \text{ Cal}}{1000 \text{ cal}} = 0.249 \text{ Cal}$

c) $113 \text{ cal} \times \frac{4.184 \text{ J}}{1 \text{ cal}} \times \frac{1 \text{ kWh}}{3.60\text{x}10^6 \text{ J}} = 1.31\text{x}10^{-4} \text{ kWh}$

d) $44 \text{ kJ} \times \frac{1000 \text{ J}}{1 \text{ kJ}} \times \frac{1 \text{ cal}}{4.184 \text{ J}} = 1.1\text{x}10^4 \text{ cal}$

55.

J	cal	Cal	kWh
225 J	53.8 cal	5.38×10^{-2} Cal	6.25×10^{-5} kWh
3.44×10^{6}J	8.21×10^{5} cal	8.21×10^{2} Cal	9.54×10^{-1} kWh
1.06×10^{9} J	2.54×10^{8} cal	2.54×10^{5} Cal	295 kWh
6.49×10^{5} J	1.55×10^{5} cal	155 Cal	1.80×10^{-1} kWh

57. $1027\ \text{kWh} \times \dfrac{3.60x10^6\ \text{J}}{1\ \text{kWh}} = 3.697x10^9\ \text{J}$

59. $2.2x10^3$ Cal - $2.0x10^3$ Cal = $2x10^2$ Cal

$$\frac{2x10^2\ \text{Cal}}{\text{day}} x \frac{4.184\ \text{J}}{\text{cal}} x \frac{1000\ \text{cal}}{1\ \text{Cal}} x \frac{1\ \text{kJ}}{1000\ \text{J}} = 8x10^2\ \text{kJ}$$

$$\frac{14.6x10^3\ \text{kJ}}{1\ \text{lb.}} x \frac{1\ \text{day}}{8.368x10^2\ \text{kJ}} = 20\ \text{days}$$

Energy and Chemical and Physical Change

61. The reaction of iron with oxygen in the atmosphere releases heat, therefore it is an exothermic reaction. See Figure 3.16a.

63. a) exothermic, $-\Delta H$
b) endothermic, $+\Delta H$
c) exothermic, $-\Delta H$

Converting Between Temperature Scales

65. a) $°\text{C} = \dfrac{[°\text{F} - 32]}{1.8} = \dfrac{[212 - 32]}{1.8} = \dfrac{180}{1.8} = 1.00x10^2\ °\text{C}$

b) $°\text{C} = \text{K} - 273 = 77 - 273 = -196°\text{C}$

$°\text{F} = (1.8 \times °\text{C}) + 32 = (1.8\ x - 196) + 32 = -3.2x10^2\ °\text{F}$

c) $\text{K} = 273 + °\text{C} = 273 + 25 = 298\ \text{K}$

d) $°\text{C} = \dfrac{[°\text{F} - 32]}{1.8} = \dfrac{[98.6 - 32]}{1.8} = 37°\text{C};\quad \text{K} = 273 + °\text{C} = 3.10x10^2\ \text{K}$

67. $°\text{C} = \dfrac{[°\text{F} - 32]}{1.8} = \dfrac{[-80 - 32]}{1.8} = \dfrac{-112}{1.8} = -62°\text{C}$

$\text{K} = 273 + °\text{C} = 273 + -62 = 211\ \text{K}$

69. $°\text{F} = (1.8 \times °\text{C}) + 32 = (1.8 \times -114) + 32 = -205.2 + 32 = -173°\text{F}$

$\text{K} = °\text{C} + 273 = -114 + 273 = 159\ \text{K}$

71. $°\text{F} = (1.8 \times -59.7°\text{C}) + 32 = -75.5°\text{C}$

73.

Kelvin	Fahrenheit	Celsius
0.0	-459 °F	-273 °C
301 K	82.5 °F	28.1 °C
282 K	47 °F	8.5 °C

75. $q = m \times c \times \Delta T$

$$q = 65\ \text{g} \times 4.184 \frac{\text{J}}{\text{g} \cdot {}^\circ\text{C}} \times (65^\circ\text{C} - 32^\circ\text{C}) = 9.0\text{x}10^3\text{J}$$

77. $$45\ \text{kg} \times \frac{1000\ \text{g}}{1\ \text{kg}} \times \frac{2.42\ \text{J}}{\text{g}\ {}^\circ\text{C}} \times (19^\circ\text{C} - 11^\circ\text{C}) = 8.7\text{x}10^5\text{J}$$

79. $$89\ \text{J} = 12\ \text{g} \times \frac{0.128\ \text{J}}{\text{g}\ {}^\circ\text{C}} \times (\Delta\text{T}) \Rightarrow \Delta\text{T} = \frac{89\ \text{J}}{12\ \text{g} \times \frac{0.128\ \text{J}}{\text{g}\ {}^\circ\text{C}}} = 58^\circ\text{C}$$

81. $$15\ \text{J} = 12\ \text{g} \times \frac{0.449\ \text{J}}{\text{g}\ {}^\circ\text{C}} \times (\text{T}_\text{f} - 28^\circ\text{C}) \Rightarrow (\text{T}_\text{f} - 28^\circ\text{C}) = \frac{15\ \text{J}}{12\ \text{g} \times \frac{0.449\ \text{J}}{\text{g}\ {}^\circ\text{C}}} \Rightarrow$$

$$\text{T}_\text{f} = 2.78 + 28^\circ\text{C} = 31\ {}^\circ\text{C}$$

83. $$248\ \text{cal} \times \frac{4.184\ \text{J}}{1\ \text{cal}} = 24\ \text{g} \times \frac{4.18\ \text{J}}{\text{g}\ {}^\circ\text{C}} \times (\Delta\text{T}) \Rightarrow \Delta\text{T} = \frac{1037.6}{100.3} = 1.0\text{x}10^1\ {}^\circ\text{C}$$

85. $58\ \text{J} = 28\ \text{g} \times (\text{heat capacity}) \times (39.9 - 31.1)^\circ\text{C} \Rightarrow$

$$\text{heat capacity} = \frac{58\ \text{J}}{(28\ \text{g})(8.8^\circ\text{C})} = 0.24 \frac{\text{J}}{\text{g}\ {}^\circ\text{C}}$$

$\therefore$ It is consistent with silver metal.

87. $56\ \text{J} = 11\ \text{g} \times (\text{heat capacity}) \times (12.7\text{-}10.4)^\circ\text{C} \Rightarrow$

$$\text{heat capacity} = \frac{56\ \text{J}}{(11\ \text{g})(2.3^\circ\text{C})} = 2.2 \frac{\text{J}}{\text{g}\ {}^\circ\text{C}}$$

89. When warm drinks are placed into the ice they release heat, which then melts the ice. The pre-chilled drinks, on the other hand, are already cold so they do not release much heat and do not melt the ice.

Cumulative Problems

91. $17\text{ kJ} \times \frac{1000\text{ J}}{1\text{ kJ}} = 245\text{g} \times \frac{4.18\text{ J}}{\text{g }^\circ\text{C}} \times (T_f - 32^\circ\text{C}) \Rightarrow$

$$(T_f - 32^\circ\text{C}) = \frac{1.7\text{x}10^4\text{ J}}{245\text{g} \times \frac{4.18\text{ J}}{\text{g }^\circ\text{C}}} \Rightarrow T_f = 16.6 + 32^\circ\text{C} = 49\ ^\circ\text{C}$$

* Assume 1 mL of water is equal to 1 gram.

93. $\text{Heat} = 1.57\text{ cm}^3 \times \frac{19.3\text{ g}}{1\text{ cm}^3} \times \frac{0.128\text{ J}}{\text{g }^\circ\text{C}} \times (29.5 - 11.4^\circ\text{C}) = 70.2\text{ J}$

95. $\text{kJ} = 56\text{ L} \times \frac{1000\text{ mL}}{1\text{ L}} \times \frac{1\text{ g}}{1\text{ mL}} \times \frac{4.18\text{ J}}{\text{g }^\circ\text{C}} \times 71^\circ\text{C}* \times \frac{1\text{ kJ}}{1000\text{ J}} = 1.7\text{x}10^4\text{ kJ}$

**You must convert each temperature prior to taking the difference.*

$T_f = (212^\circ\text{F} - 32) \times \frac{5}{9} = 100^\circ\text{C}; \quad T_I = (85^\circ\text{F} - 32) \times \frac{5}{9} = 29^\circ\text{C}$

$\Delta\text{T} = 100^\circ\text{C} - 29^\circ\text{C} = 71^\circ\text{C}$

97. $2.3\text{ kWh} \times \frac{3.60\text{x}10^6\text{J}}{1\text{ kWh}} = 29.5\text{ L} \times \frac{1000\text{ mL}}{1\text{ L}} \times \frac{1\text{ g}}{1\text{ ml}} \times \frac{4.18\text{ J}}{\text{g }^\circ\text{C}} \times \Delta\text{T} \Rightarrow$

$$\Delta\text{T} = \frac{2.3\text{ kWh} \times \frac{3.60\text{x}10^6\text{J}}{1\text{ kWh}}}{29.5\text{ L} \times \frac{1000\text{ mL}}{1\text{ L}} \times \frac{1\text{ g}}{1\text{ ml}} \times \frac{4.18\text{ J}}{\text{g }^\circ\text{C}}} = \frac{8.28\text{x}10^6}{1.045\text{x}10^5} = 67^\circ\text{C}$$

99. $55\text{ gal} \times \frac{3.785\text{L}}{1\text{ gal}} \times \frac{1000\text{mL}}{1\text{ L}} \times \frac{1\text{ g}}{1\text{ mL}} \times \frac{4.18\text{J}}{\text{g }^\circ\text{C}} \times 25^\circ\text{C} \times \frac{1\text{ kWh}}{3.60\text{x}10^6\text{J}} = 6.0\text{ kWh}$

101. $2.5\text{kg H}_2\text{O} \times \frac{1000\text{ g}}{1\text{ kg}} \times \frac{4.184\text{ J}}{\text{g }^\circ\text{C}} \times 75^\circ\text{C} \times \frac{1\text{ kJ}}{1000\text{ J}} \times \frac{1\text{ g Fuel}}{36\text{ kJ}} = 22\text{g Fuel}$

103. $95\text{kg} \times \frac{1000\text{ g}}{1\text{ kg}} \times \frac{4.0\text{ J}}{\text{g }^\circ\text{C}} \times 0.50^\circ\text{C} \times \frac{1\text{ kJ}}{1000\text{ J}} \times \frac{1\text{ g H}_2\text{O}}{2.44\text{ kJ}} = 78\text{ g H}_2\text{O}$

105. Heat lost by Aluminum = Heat gained by water

$$15.7\text{g Al}\times\frac{0.903\text{J}}{\text{g °C}}\times(53.2\text{°C-T}_f)=32.5\text{g H}_2\text{O}\times\frac{4.184\text{J}}{\text{g °C}}\times(\text{T}_f\text{-24.5°C})\Rightarrow$$

$$\frac{14.\underline{1}8\text{J}}{\text{°C}}\times(53.2\text{°C-T}_f)=\frac{13\underline{6}.0\text{ J}}{\text{°C}}\times(\text{T}_f\text{-24.5°C})\Rightarrow$$

$$75\underline{4}\text{ J-}\frac{14.\underline{1}8\text{ J}}{\text{°C}}(\text{T}_f)=\frac{13\underline{5}.98\text{ J}}{\text{°C}}(\text{T}_f)\text{-33}\underline{3}\text{2 J}\Rightarrow$$

$$(75\underline{4}\text{ J+33}\underline{3}\text{2 J})=\left(\frac{13\underline{6}.0\text{ J}}{\text{°C}}+\frac{14.\underline{1}8\text{ J}}{\text{°C}}\right)(\text{T}_f)$$

$$40\underline{8}6\text{ J}=\left(\frac{15\underline{0}.1\text{ J}}{\text{°C}}\right)(\text{T}_f)\Rightarrow \text{T}_f=\frac{40\underline{8}6\text{ J}}{15\underline{0}.1\text{ J/°C}}=27.2\text{ °C}$$

107. Watts → J/s → kJ/month; kJ/month → $/month

855W - 625W = 230W = 230J/s

$$\frac{230\text{ J}}{\text{s}}\times\frac{60\text{s}}{1\text{min}}\times\frac{60\text{min}}{1\text{ hr}}\times\frac{24\text{ hr}}{1\text{ day}}\times\frac{30\text{ day}}{1\text{ month}}\times\frac{1\text{ kJ}}{1000\text{ J}}=5.96x10^5\frac{\text{kJ}}{\text{month}}$$

$$5.96x10^5\frac{\text{kJ}}{\text{month}}\times\frac{1\text{ kWh}}{3.60\text{x}10^3\text{kJ}}\times\frac{\$0.15}{\text{kWh}}=\$24.83$$

109. Set °C = °F in the equation to convert temperatures between scales, solve.
°C 9/5 + 32 = °F → 1.8x + 32 = x →1.8x - x = -32 →0.8x = -32 → x = -32/0.8 = -40
-40 °C = -40 °F

111. a) pure substance
b) pure substance
c) pure substance
d) mixture

113. physical change

Highlight Problems

115. The weather patterns that develop over the ocean exchange heat with the water. If the water is several degrees colder than the air above it, an extremely large amount of heat is removed from the air to the water. This is because water has a large heat capacity and air has a low heat capacity. Conversely, if the water is several degrees warmer than the air above it, a tremendous amount of heat can be transferred to the air. This occurs all of the time and is part of our normal weather patterns. However, during the El Niño/La Niña cycle, this heat transfer to and from water and air increases dramatically, which is why it alters weather patterns on a global scale.

117. a) San Francisco is nearly surrounded by the ocean. Because water can absorb large amounts of heat without an increase in temperature (i.e., water has a large heat capacity), San Francisco enjoys moderate temperatures. Sacramento, on the other hand, is a land-locked city. The earth has a relatively low heat capacity, which means that as the earth absorbs heat, the temperature quickly increases (i.e., a small heat capacity).

    b) In the winter, San Francisco is a warmer place. The ocean releases a large amount of heat back into the colder atmosphere because it can store a tremendous amount of heat. Sacramento, on the other hand, is surrounded by land and is cooler because the earth's heat capacity is much lower than the ocean's.

Atoms and Elements

4

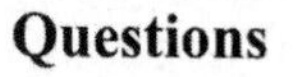

Questions

1. Democritus reasoned that matter was made of small, indivisible, indestructible particles.

3. Rutherford had shot alpha particles at an extremely thin gold foil target. The majority of alpha particles passed directly through the foil, some particles were slightly deflected and 1 in 20,000 bounced back toward the alpha source. If the plum pudding model had been correct, the only result would have been a slight deflection of some of the particles.

5.

Particle	Mass (kg)	Mass (amu)	Charge	Location
Proton	1.67262×10^{-27}	1	+1	Inside Nucleus
Neutron	1.67493×10^{-27}	1	0	Inside Nucleus
Electron	0.00091×10^{-27}	0.00055	-1	Outside Nucleus

7. Matter is usually charge neutral because the protons and electrons cancel each other and, when an imbalance does exist, it usually corrects itself very soon.

9. A chemical symbol is a one or two letter abbreviation that is unique to each element.

11. Mendeleev was the first person to organize the elements into what we would recognize as a version of the modern periodic table. Mendeleev based his table on increasing molecular mass of elements and grouping elements with similar properties into columns.

13. The modern periodic table is organized by increasing atomic number, which is the number of protons found in each element.

15. The properties of nonmetals are: (1) poor conductors of heat and electricity; (2) tend to gain electrons in reactions; (3) can be found as solids, liquids, or gases. Nonmetals make up the upper right side of the periodic table.

17. A family or group of elements is the term given to the column on a periodic table which contains elements that all have similar, periodic properties.

19. An ion is an atom (or group of atoms) that becomes electrically charged due to a gain or loss of electrons.

21. a) +1 b) +2 c) +3 d) -2 e) -1

23. The percent natural abundance of isotopes provides the relative amounts of each isotope found in a sample of the element. These numbers are constant no matter the source of the element and are unique to each different element.

25. The first method for specifying isotopes is to superscript and subscript the mass number and atomic number, respectively, on the left side of the chemical symbol (e.g., ${}^{2}_{1}H$). The second method is to identify the element followed by the mass number (e.g., H-2 or hydrogen-2).

Problems

Atomic and Nuclear Theory

27. a) consistent; all atoms of a given element have the same mass and other properties that distinguish it from the atoms of other elements.
    b) inconsistent; each element is composed of tiny indestructible particles called atoms.
    c) inconsistent; atoms combine in simple, whole-number ratios to form compounds.
    d) consistent; atoms combine in simple, whole-number ratios to form compounds.

29. a) consistent; there are as many negatively charged electrons outside the nucleus as there are positively charged particles (called protons) inside the nucleus, therefore the atom is electrically neutral.
    b) inconsistent; most of the volume of the atom is empty space occupied by tiny, negatively charged electrons.
    c) inconsistent; there are as many negatively charged electrons outside the nucleus as there are positively charged particles (called protons) inside the nucleus, therefore the atom is electrically neutral.
    d) inconsistent; most of the atom's mass and all of its positive charge are contained in a small core called the nucleus.

31. Matter appears solid because the variation in the density is on such a small scale that our eyes cannot distinguish the difference.

Protons, Neutrons, and Electrons

33. a) True
    b) True
    c) False; all electrons have the same mass and charge.
    d) True

35. a) False; protons and neutrons are nearly identical in mass.
    b) True
    c) False; neutral atoms have an equal number of protons and electrons.
    d) True

37. (Mass of 1 Electron)(X) = (Mass of 1 Proton) $\Rightarrow$

$$X=\frac{\text{(Mass of 1 Proton)}}{\text{(Mass of 1 Electron)}}=\frac{1.67262\times10^{-27}\text{kg}}{0.00091\times10^{-27}\text{kg}}=1.8\times10^{3}$$

Elements, Symbols, and Names

39. $$1.0 \text{ g protons}\times\frac{1 \text{ proton}}{1.67262\times10^{-24}\text{g}}\times\frac{1 \text{ electron}}{1 \text{ proton}}\times\frac{9.1\times10^{-28}\text{g}}{1 \text{ electron}}=5.4\times10^{-4}\text{g}$$

41. a) 87
    b) 36
    c) 91
    d) 32
    e) 13

43. a) 18
    b) 50
    c) 54
    d) 8
    e) 81

45. a) C, 6
    b) N, 7
    c) Na, 11
    d) K, 19
    e) Cu, 29

47. a) Manganese, 25
    b) Silver, 47
    c) Gold, 79
    d) Lead, 82
    e) Sulfur, 16

49.

Element Name	Element Symbol	Atomic Number
Gold	Au	79
Tin	Sn	50
Arsenic	As	33
Copper	Cu	29
Iron	Fe	26
Mercury	Hg	80

The Periodic Table

51. a) metal b) metal c) nonmetal d) nonmetal e) metalloid

53. Metals lose electrons in reactions therefore a) potassium, d) barium, and e) copper

55. a) Te and b) K

57. c) calcium and d) barium

59. b) sodium and e) rubidium

61. a) halogen b) noble gas c) halogen d) neither e) noble gas

63. a) 16 or VIA b) 13 or IIIA c)14 or IVA d) 14 or IVA e) 15 or VA

65. The elements most like sulfur would be those elements in the same group (6A) because they all have similar physical and chemical properties. Therefore, the answer is b) oxygen.

67. The elements that come from the same group would be the most similar because they have similar chemical and physical properties. Therefore, the answer is b) Cl and F.

69.

Chemical Symbol	Group Number	Group Name	Metal or Nonmetal
K	1A or 1	alkali metal	metal
Br	7A or 17	halogen	nonmetal
Sr	2A or 2	alkaline earth	metal
He	8A	noble gas	nonmetal
Ar	8A or 18	noble gas	nonmetal

Ions

71. a) $Na \rightarrow Na^{+} + e^{-}$
b) $O + 2\ e^{-} \rightarrow O^{2-}$
c) $Ca \rightarrow Ca^{2+} + 2e^{-}$
d) $Cl + e^{-} \rightarrow Cl^{-}$

73. a) oxygen ion charge = 8 (+1) + 10 (-1) = -2
b) aluminum ion charge = 13 (+1) + 10 (-1) = +3
c) titanium ion charge = 22 (+1) + 18 (-1) = +4
d) iodine ion charge = 53 (+1) + 54 (-1) = -1

75. The number of protons is determined using the atomic number of each element. The number of electrons is determined by examining the net charge on the ion.
   a) 11p + 10e- = +1
   b) 56p + 54e- = +2
   c) 8p + 10e- = -2
   d) 27p + 24e- = +3

77. a) False; The Ti^{2+} ion contains 22 protons and 20 electrons.
   b) True
   c) False; The Mg^{2+} ion contains 12 protons and 10 electrons.
   d) True

79. a) Rb is in group 1A, therefore it will form Rb^{+}.
   b) K is in group 1A, therefore it will form K^{+}.
   c) Al in group 3A, therefore it will form Al^{3+}.
   d) O is in group 6A, therefore it will form O^{2-}.

81. a) Ga is in group 3A, therefore it will lose 3 electrons.
   b) Li is in group 1A, therefore it will lose 1 electron.
   c) Br is in group 7A, therefore it will gain 1 electron.
   d) S is in group 6A, therefore it will gain 2 electrons.

83.

Symbol	Ion Formed	# electrons in ion	# protons in ion
Te	Te^{2-}	54	52
In	In^{3+}	46	49
Sr	Sr^{2+}	36	38
Mg	Mg^{2+}	10	12
Cl	Cl^{-}	18	17

Isotopes

85. a) Z=1, A=3
   b) Z=24, A=52
   c) Z=20, A=42
   d) Z=73, A=182

87. a) ${}^{16}_{8}O$
b) ${}^{19}_{9}F$
c) ${}^{23}_{11}Na$
d) ${}^{27}_{13}Al$

89. a) ${}^{60}_{27}Co$
b) ${}^{22}_{10}Ne$
c) ${}^{131}_{53}I$
d) ${}^{244}_{94}Pu$

91. a) protons = Z = 11, neutrons = A–Z = 23–11 = 12
b) protons = Z = 88, neutrons = A–Z = 266–88 = 178
c) protons = Z = 82, neutrons = A–Z = 208–82 = 126
d) protons = Z = 7, neutrons = A–Z = 14–7 = 7

93. Carbon-14 has Z=6, therefore, protons = Z = 6, and neutrons = A–Z= 14–6 = 8. The correct isotope symbol would be ${}^{14}_{6}C$.

Atomic Mass

95. Calculating the atomic mass of an element involves taking a weighted average in which you multiply the percent natural abundance by the atomic mass for each isotope and then add these products together. For rubidium:
Atomic Mass = Σ(isotopic abundance) × (isotopic mass) ⇒
(0.7217)(84.9118 amu) + (0.2783)(86.9092 amu) = 85.47 amu

97. a) 100–50.69 = 49.31%
b) (0.5069)(mass) + (0.4931)(80.9163) = 79.904 amu ⇒
Mass = (79.904–39.90)/0.5069 ⇒
Mass = 40.00/0.5069 = 78.92 amu

99. MW = (0.574)(120.9038)+(0.426)(122.9042) ⇒ MW = 69.4 + 52.4 = 121.8 amu
The atomic weight is closest to that of antimony (Sb) which is 121.75 amu.

Cumulative Problems

101. $-125\text{ mC} \times \frac{1\text{ C}}{1000\text{ mC}} \times \frac{1\text{ e}^-}{-1.6\times10^{-19}\text{C}} = 7.8\times10^{17}\text{e}^-$

103. Nucleus: $V=\frac{4}{3}\pi(1.0\times10^{-15}\text{m})^3 = 4.2\times10^{-45}\text{m}^3$

Hydrogen: $V=\frac{4}{3}\pi(53\times10^{-12}\text{m})^3 = 6.2\times10^{-31}\text{m}^3$

Percentage: $\frac{4.2\times10^{-45}\text{m}^3}{6.2\times10^{-31}\text{m}^3}\times100 = 6.8\times10^{-13}\%$

105.

Symbol	#p	#n	A (Mass Number)	Natural Abundance
Sr-84 or $^{84}_{38}Sr$	38	46	84	0.56%
Sr-86 or $^{86}_{38}Sr$	38	48	86	9.86%
Sr-87 or $^{87}_{38}Sr$	38	49	87	7.00%
Sr-88 or $^{88}_{38}Sr$	38	50	88	82.58%

Atomic Mass of Sr is the weighted average of each isotope:

$Sr = (0.0056\times83.9134)+(0.0986\times85.9093)+(0.0700\times86.9089)+(0.8258\times87.9056)$

$Sr = 0.47+8.47+6.08+72.59 = 87.61$ amu

107.

Symbol	Z	A	#p	#e$^-$	#n	Charge
Zn^+	30	64	30	29	34	1+
Mn^{3+}	25	55	25	22	30	3+
P	15	31	15	15	16	0
O^{2-}	8	16	8	10	8	2$^-$
S^{2-}	16	34	16	18	18	2$^-$

109. % abundance of Eu-153: 100 – 47.8 = 52.2%
(0.478)(150.9198 amu) + (0.522)(Mass ^{153}Eu) = 151.97 amu ⇒
Mass ^{153}Eu = (151.97–72.1397)/(0.522) =153 amu

111. The atomic theory and nuclear model of the atom are both theories because they attempt to provide a broader understanding and model behavior of chemical systems.

113. The atomic mass reported on the periodic table is a weighted average of all natural stable isotopes. As fluorine only has one isotope, the atomic mass is identical to the mass of the isotope. Chlorine, however, must have more than one stable isotope that occurs naturally. The relative abundance of each isotope factors into creating the average atomic mass reported on the periodic table.

115. Set x= abundance of Cu-63, then 1–x = abundance of Cu-65

$(62.9396)(x) + (64.9278)(1-x) = 63.55$

$62.9396x + 64.9278 - 64.9278x = 63.55$

$62.9396x - 64.9278x = 63.55 - 64.9278$

$-1.9882x = -1.38 \quad x = -1.38/-1.9882 = 0.693$ (69.3%)

Cu-65 = 1–0.693 = 0.307 (30.7%)

Highlight Problems

117. a) Nt-304: $\frac{36}{50}\times 100\% = 72\%$; Nt-305: $\frac{2}{50}\times 100\% = 4.0\%$; Nt-306: $\frac{12}{50}\times 100\% = 24\%$

b) MW of Nt = (0.72)(303.956)+(0.040)(304.962)+(0.24)(305.978)

=304.5 amu (assuming the percentages have 4 significant digits)

120
Nt
304.5

Molecules and Compounds 5

Questions

1. The properties of an element completely change when it combines with another element to form a compound. For example, water is made out of the elements hydrogen and oxygen, both of which are gases. When hydrogen and oxygen combine to form water, their properties change and they become a new liquid substance.

3. The law of constant composition was first expressed by Joseph Proust and states: All samples of a given compound have the same proportions of their constituent elements.

5. The general rule for listing elements in a compound is that the most metallic element is listed first. For compounds which contain a metal, it is listed first. For compounds that do not contain a metal, the most metal-like element is listed first. For nonmetals, you list the element that is found farthest to the left and/or the element that is lowest on the periodic table.

7. An empirical formula gives the simplest whole number ratio of atoms of each element in a compound whereas a molecular formula gives the actual number of atoms of each element in a compound.

9. Most elements can be found in nature as single atoms, which we call the **atomic elements**. However, some elements can only be found in nature in the diatomic state, that is two atoms of the same element bonded together which we call the **molecular elements**. There are only 7 diatomic molecular elements: hydrogen (H_2), nitrogen (N_2), oxygen (O_2), fluorine (F_2), chlorine (Cl_2), bromine (Br_2), and iodine (I_2). *It is tradition that when a chemist says a molecular element by name, they are referring to the diatomic state. When your instructor says oxygen reacts with hydrogen, you know to write O_2 and H_2. For a reaction that involves just a single atom of these elements you would say monatomic oxygen (O).*

11. The systematic name for a compound will provide the reader with enough information that the formula of the compound can be determined. The common name can be considered a "nickname" for the compound that must be memorized.

13. The metals that form Type II Ionic Compounds are most commonly found in the center section of the periodic table, known as the transition metals.

15. The basic form for naming Type II Binary compounds is the same as Type I with the addition of the charge of the cation, written in roman numerals, inserted in parentheses between the cation name and the anion name. The pattern is: [Cation Name](cation charge-roman numerals) [Anion Base Name + *-ide*]

17. When you name compounds that contain polyatomic ions, you use the same rules from Type I and Type II, however, you insert the name of the polyatomic ions, without altering the name, into their proper place.

19. The form for naming molecular compounds is to name the more metallic element (i.e., the left-most element in the periodic table) first. The less metallic element (i.e., right-most in the periodic table) is named second, adding the -ide suffix. Each element name is preceded by a numerical prefix to indicate the number of each atom in the compound.

21. The basic form for naming binary acids is: [hydro + base nonmetal name + "ic"][acid].

23. The basic form for naming oxyanions with the "-ite" ending is: [base oxyanion name + "ous"][acid].

Problems

Constant Composition of Compounds

25. Sample 1: $\frac{\text{mass Cl}}{\text{mass Na}} = \frac{7.16}{4.65} = 1.54$

Sample 2: $\frac{\text{mass Cl}}{\text{mass Na}} = \frac{11.5}{7.45} = 1.54$, Yes

27. Sample 1: $\frac{\text{mass F}}{\text{mass Mg}} = \frac{2.57}{1.65} = 1.56$

Sample 2: $\frac{\text{mass F}}{\text{mass Mg}} = \frac{\text{mass F}}{1.32} = 1.56$

mass F $= 1.56 \times 1.32 = 2.06$ kg

29.

	Mass N_2O	Mass N	Mass O
Sample A	2.85	1.82	1.04
Sample B	4.55	<u>2.91</u>	<u>1.66</u>
Sample C	<u>3.70</u>	<u>2.36</u>	1.35
Sample D	<u>1.74</u>	1.11	<u>0.635</u>

Chemical Formulas

31. NI_3

33. a) Fe_3O_4
    b) PCl_3
    c) PCl_5
    d) Ag_2O

35. a) 4
    b) 4
    c) 2×3=6
    d) 2×2=4

37. a) Mg=1, Cl=2
    b) Na=1, N=1, O=3
    c) Ca=1, N=2×1=2, O=2×2=4
    d) Sr=1, O=2×1=2, H=2×1=2

39. Complete the following table

Formula	Number of $C_2H_3O_2^-$ units	No. of C atoms	No. of H atoms	No. of O atoms	No. of metal atoms
$Mg(C_2H_3O_2)_2$	2	4	6	4	1
$NaC_2H_3O_2$	1	2	3	2	1
$Cr_2(C_2H_3O_2)_4$	4	8	12	8	2

41. a) CH_3
    b) NO_2
    c) C_2H_3O
    d) NH_3

Molecular View of Elements and Compounds

43. a) molecular
    b) atomic
    c) atomic
    d) molecular

45. a) molecular
    b) ionic
    c) ionic
    d) molecular

47. Helium — single atoms
    CCl_4 — molecules
    K_2SO_4 — formula units
    bromine — diatomic molecules

49. a) formula units
    b) single atoms
    c) molecules
    d) molecules

51. a) ionic, single ion type
    b) molecular
    c) molecular
    d) ionic, multiple ion types

Writing Formulas for Ionic Compounds

53. a) Na =+1, S =−2: Na_2S
    b) Sr=+2, O=−2: SrO
    c) Al=+3, S=−2: Al_2S_3
    d) Mg=+2, Cl=−1: $MgCl_2$

55. a) $KC_2H_3O_2$
    b) K_2CrO_4
    c) K_3PO_4
    d) KCN

57. N=-3, O=-2, F=-1
    a) Li=+1: Li_3N, Li_2O, LiF
    b) Ba=+2: Ba_3N_2, BaO, BaF_2
    c) Al=+3: AlN, Al_2O_3, AlF_3

Naming Ionic Compounds

59. a) cesium chloride
    b) strontium bromide
    c) potassium oxide
    d) lithium fluoride

61. a) chromium (II) chloride
    b) chromium (III) chloride
    c) tin (IV) oxide
    d) lead (II) iodide

63. a) multiple ion types
    b) single ion type
    c) single ion type
    d) multiple ion types

65. a) barium nitrate
    b) lead (II) acetate
    c) ammonium iodide
    d) potassium chlorate
    e) cobalt (II) sulfate
    f) sodium perchlorate

67. a) hypobromite ion
    b) bromite ion
    c) bromate ion
    d) perbromate ion

69. a) $CuBr_2$
    b) $AgNO_3$
    c) KOH
    d) Na_2SO_4
    e) $KHSO_4$
    f) $NaHCO_3$

Naming Molecular Compounds

71. a) sulfur dioxide
    b) nitrogen triiodide
    c) bromine pentafluoride
    d) nitrogen monoxide
    e) tetranitrogen tetraselenide

73. a) CO
    b) S_2F_4
    c) Cl_2O
    d) PF_5
    e) BBr_3
    f) P_2S_5

75. a) incorrect, phosphorus pentabromide
    b) incorrect, diphosphorus trioxide
    c) incorrrect, sulfur tetrafluoride
    d) correct

Naming Acids

77. a) oxyacid, nitrous acid, nitrite ion
    b) binary, hydroiodic acid
    c) oxyacid, sulfuric acid, sulfate ion
    d) oxyacid, nitric acid, nitrate ion

79. a) hypochlorous acid
    b) chlorous acid
    c) chloric acid
    d) perchloric acid

81. a) H_3PO_4
    b) HBr (*aq*)
    c) H_2SO_3

Formula Mass

83. a) FM=1.01+14.01+3(16.00)=63.02 amu
    b) FM=40.08+2(79.90)=199.88 amu
    c) FM=12.01+4(35.45)=153.81 amu
    d) FM=87.62+2(14.01)+6(16.00)=211.64 amu

85. PBr_3(FM=270.67 amu) > Ag_2O(FM=231.74 amu) > PtO_2(FM=227.08 amu) > $Al(NO_3)_3$(FM=213.01 amu)

Cumulative Problems

87. a) CH_4
    b) SO_3
    c) NO_2

89. a) 3×4=12
    b) 2×2=4
    c) 4×3=12
    d) 7×1=7

91. a) 8
    b) 12
    c) 12

93.

Formula	Type	Name
N_2H_4	molecular	dinitrogen tetrahydride
KCl	ionic	potassium chloride
H_2CrO_4(*aq*)	acid	chromic acid
$Co(CN)_3$	ionic	cobalt(III) cyanide

95. a) incorrect, calcium nitrite
    b) incorrect, potassium oxide
    c) incorrect, phosphorus trichloride
    d) correct
    e) potassium iodite

97. a) $Sn(SO_4)_2$, FM=118.71+2(32.07)+8(16.00)=310.85 amu
b) HNO_2, FM=1.008+14.01+2(16.00)=47.02 amu
c) $NaHCO_3$, FM=22.99+1.008+12.01+3(16.00)=84.01 amu
d) PF_5, FM=30.97+5(19.00)=125.97 amu

99. a) platinum (IV) oxide, FM=195.08+2(16.00)=227.08 amu
b) dinitrogen pentoxide, FM=2(14.01)+5(16.00)=108.02 amu
c) aluminum chlorate, FM=26.98+3(35.45)+9(16.00)=277.33 amu
d) phosphorus pentabromide, FM=30.97+5(79.90)=430.47 amu

101. C=12.01g/mol, H=1.01g/mol and the compound has a formula mass of 28.06 amu. If the compound had 1 carbon atom, the remaining mass would be equal to 15.89 hydrogen atoms. If the compound has 2 carbon atoms, the remaining mass is equal to exactly 4 hydrogen atoms, therefore the formula of the compound is C_2H_4.

103. Ten different masses of CCl_4 masses, as shown in the table.

C Isotope	No. of Cl-35	No. of Cl-37	Formula Mass
C-12	4	0	151.88
	3	1	153.88
	2	2	155.88
	1	3	157.88
	0	4	159.88
C-13	4	0	152.88
	3	1	154.88
	2	2	156.88
	1	3	158.88
	0	4	160.88

Highlight Problems

105. a) molecular element
b) atomic element
c) ionic compound
d) molecular compound

107. Image 1: sodium hypochlorite: NaClO, sodium hydroxide: NaOH
Image 2: calcium carbonate: $CaCO_3$
Image 3: aluminum hydroxide: $Al(OH)_3$, magnesium hydroxide: $Mg(OH)_2$
Image 4: sodium bicarbonate: $NaHCO_3$, calcium phosphate: $Ca_3(PO_4)_2$,
sodium aluminum sulfate: $NaAl(SO_4)_2$

Chemical Composition 6

Questions

1. Chemical composition is important to understand because it provides an analysis of the amount of an element found within a compound.

3. There are 6.022×10^{23} atoms in 1 mole of atoms.

5. One mole of atoms of an element has a mass in grams that is equal to the mass of the atom in atomic mass units (amu).

7. a) 30.97 g b) 195.08 g c) 12.01 g d) 52.00 g

9. The subscripts provide a ratio of atoms of one element to atoms of another element within a compound. This ratio does not apply to mass because every element has a different mass. A one-to-one atomic ratio for hydrogen to oxygen is very different than the one-to-one atomic ratio for hydrogen to sulfur because sulfur has a mass of over twice the mass of oxygen.

11. a) 11.19 g H $\equiv$ 100 g water
    b) 53.29 g O $\equiv$ 100 g fructose
    c) 84.12 g C $\equiv$ 100 g gasoline
    d) 52.14 g C $\equiv$ 100 g ethanol

13. The molecular formula is a whole-number multiple of the empirical formula.

15. The empirical formula mass is the sum of the masses of all atoms found in the simplest whole-number ratio of each atom type found in a compound.

Problems

The Mole Concept

17. moles $\rightarrow$ atoms

$$5.8 \text{ mol Hg} \times \frac{6.022 \times 10^{23} \text{ Hg atoms}}{1 \text{ mol Hg}} = 3.5 \times 10^{24} \text{ Hg atoms}$$

19. a) $3.4 \text{ mol Cu} \times \frac{6.022 \times 10^{23} \text{ Cu atoms}}{1 \text{ mole Cu}} = 2.0 \times 10^{24} \text{Cu atoms}$

b) $9.7 \times 10^{-3} \text{mol C} \times \frac{6.022 \times 10^{23} \text{ C atoms}}{1 \text{ mole C}} = 5.8 \times 10^{21} \text{C atoms}$

c) $22.9 \text{ mol Hg} \times \frac{6.022 \times 10^{23} \text{ Hg atoms}}{1 \text{ mole Hg}} = 1.38 \times 10^{25} \text{Hg atoms}$

d) $0.215 \text{ mol Na} \times \frac{6.022 \times 10^{23} \text{ Na atoms}}{1 \text{ mole Na}} = 1.29 \times 10^{23} \text{Na atoms}$

21.

Element	Moles	Number of Atoms
Ne	0.552	3.32×10^{23}
Ar	5.40	3.25×10^{24}
Xe	1.78	1.07×10^{24}
He	1.79×10^{-4}	1.08×10^{20}

23. a) $872 \text{ sheets} \times \frac{1 \text{ dozen}}{12 \text{ sheets}} = 72.7 \text{ dozen}$

b) $872 \text{ sheets} \times \frac{1 \text{ gross}}{144 \text{ sheets}} = 6.06 \text{ gross}$

c) $872 \text{ sheets} \times \frac{1 \text{ ream}}{500 \text{ sheets}} = 1.74 \text{ reams}$

d) $872 \text{ sheets} \times \frac{1 \text{ mole}}{6.022 \times 10^{23} \text{ sheets}} = 1.45 \times 10^{-21} \text{ moles}$

25. $38.1 \text{ g Sn} \times \frac{1 \text{ mole Sn}}{118.71 \text{ g Sn}} = 0.357 \text{ mole Sn}$

27. $0.145 \text{ mol Au} \times \frac{196.97 \text{ g Au}}{1 \text{ mol Au}} = 28.6 \text{ g Au}$

29. a) $1.34 \text{ g Zn} \times \frac{1 \text{ mol Zn}}{65.39 \text{ g Zn}} = 2.05 \times 10^{-2} \text{ mol Zn}$

b) $24.9 \text{ g Ar} \times \frac{1 \text{ mol Ar}}{39.95 \text{ g Ar}} = 0.623 \text{ mol Ar}$

c) $72.5 \text{ g Ta} \times \frac{1 \text{ mol Ta}}{180.95 \text{ g Ta}} = 0.401 \text{ mol Ta}$

d) $0.0223 \text{ g Li} \times \frac{1 \text{ mol Li}}{6.941 \text{ g Li}} = 3.21 \times 10^{-3} \text{ mol Li}$

31.

Element	Moles	Mass
Ne	1.11	22.5g
Ar	0.117	4.67g
Xe	7.62	1.00kg
He	1.44×10^{-4}	5.76×10^{-4}

33. $0.0134 \text{ mmol Ag} \times \dfrac{1\times10^{-3}\text{mole Ag}}{1 \text{ mmol Ag}} \times \dfrac{6.022\times10^{23}\text{ Ag atoms}}{1 \text{ mole Ag}} = 8.07\times10^{18}\text{ Ag atoms}$

35. $3.78 \text{ g Al} \times \dfrac{1 \text{ mole Al}}{26.98 \text{ g Al}} \times \dfrac{6.022\times10^{23}\text{ Al atoms}}{1 \text{ mole Al}} = 8.44\times10^{22}\text{ Al atoms}$

37. a) $16.9 \text{ g Sr} \times \dfrac{1 \text{ mol Sr}}{87.62 \text{ g Sr}} \times \dfrac{6.022\times10^{23}\text{Sr atoms}}{1 \text{ mol Sr}} = 1.16\times10^{23}\text{Sr atoms}$

b) $26.1 \text{ g Fe} \times \dfrac{1 \text{ mol Fe}}{55.85 \text{ g Fe}} \times \dfrac{6.022\times10^{23}\text{Fe atoms}}{1 \text{ mol Fe}} = 2.81\times10^{23}\text{Fe atoms}$

c) $8.565 \text{ g Bi} \times \dfrac{1 \text{ mol Bi}}{209.0 \text{ g Bi}} \times \dfrac{6.022\times10^{23}\text{Bi atoms}}{1 \text{ mol Bi}} = 2.46\times10^{22}\text{Bi atoms}$

d) $38.2 \text{ g P} \times \dfrac{1 \text{ mole P}}{30.97 \text{ g P}} \times \dfrac{6.022\times10^{23}\text{P atoms}}{1 \text{ mole P}} = 7.43\times10^{23}\text{P atoms}$

39. $38 \text{ mg C} \times \dfrac{1 \text{ g}}{1000 \text{ mg}} \times \dfrac{1 \text{ mol C}}{12.01 \text{ g}} \times \dfrac{6.022\times10^{23}\text{atoms}}{1 \text{ mol C}} = 1.9\times10^{21}\text{C atoms}$

41. $1.28 \text{ kg Ti} \times \dfrac{1\times10^{3}\text{g}}{1 \text{ kg}} \times \dfrac{1 \text{ mole Ti}}{47.88 \text{ g Ti}} \times \dfrac{6.022\times10^{23}\text{Ti atoms}}{1 \text{ mole Ti}} = 1.61\times10^{25}\text{He atoms}$

43.

Element	Mass	Moles	Number of Atoms
Na	38.5 mg	1.67×10^{-3}	1.01×10^{21}
C	13.5 g	1.12	6.74×10^{23}
V	1.81×10^{-20} g	3.55×10^{-22}	214
Hg	1.44 kg	7.18	4.32×10^{24}

45. a) $27.2 \text{ g Cr} \times \dfrac{1 \text{ mol Cr}}{52.00 \text{ g}} \times \dfrac{6.022\times10^{23}\text{atoms}}{1 \text{ mol Cr}} = 3.15\times10^{23}\text{Cr atoms}$

b) $55.1 \text{ g Ti} \times \dfrac{1 \text{ mol Ti}}{47.87 \text{ g}} \times \dfrac{6.022\times10^{23}\text{atoms}}{1 \text{ mol Ti}} = 6.93\times10^{23}\text{Ti atoms}$

c) $205 \text{ g Pb} \times \dfrac{1 \text{ mol Pb}}{207.2 \text{ g}} \times \dfrac{6.022\times10^{23}\text{atoms}}{1 \text{ mol Pb}} = 5.96\times10^{23}\text{Pb atoms}$

The greatest number of atoms are found in b) 55.1 g of Titanium.

47. a) $38.2 \text{ g NaCl} \times \frac{1 \text{ mol NaCl}}{58.44 \text{ g}} = 0.654 \text{ mol NaCl}$

b) $36.5 \text{ g NO} \times \frac{1 \text{ mol NO}}{30.01 \text{ g}} = 1.22 \text{ mol NO}$

c) $4.25 \text{ kg } CO_2 \times \frac{1000 \text{ g}}{1 \text{ kg}} \times \frac{1 \text{ mol } CO_2}{44.01 \text{ g}} = 96.6 \text{ mol } CO_2$

d) $2.71 \text{ mg } CCl_4 \times \frac{1 \text{ g}}{1000 \text{ mg}} \times \frac{1 \text{ mol } CCl_4}{153.8 \text{ g}} = 1.76 \times 10^{-5} \text{ mol } CCl_4$

49.

Compound	Mass	Moles	Number of Molecules
H_2O	112 kg	6.22×10^3	3.74×10^{27}
N_2O	6.33 g	0.144	8.67×10^{22}
SO_2	156 g	2.44	1.47×10^{24}
CH_2Cl_2	5.46 g	0.0643	3.87×10^{22}

51. $1.32 \text{ g } C_{10}H_8 \times \frac{1 \text{ mole } C_{10}H_8}{128.18 \text{ g } C_{10}H_8} \times \frac{6.022 \times 10^{23} C_{10}H_8}{1 \text{ mole } C_{10}H_8} = 6.20 \times 10^{21} C_{10}H_8 \text{ molecules}$

53. a) $3.5 \text{ g } H_2O \times \frac{1 \text{ mole } H_2O}{18.02 \text{ g}} \times \frac{6.022 \times 10^{23} \text{molecules}}{1 \text{ mole } H_2O} = 1.2 \times 10^{23} H_2O \text{ molecules}$

b) $56.1 \text{ g } N_2 \times \frac{1 \text{ mole } N_2}{28.02 \text{ g}} \times \frac{6.022 \times 10^{23} \text{molecules}}{1 \text{ mole } N_2} = 1.21 \times 10^{24} N_2 \text{ molecules}$

c) $89 \text{ g } CCl_4 \times \frac{1 \text{ mole } CCl_4}{153.81 \text{ g}} \times \frac{6.022 \times 10^{23} \text{molecules}}{1 \text{ mole } CCl_4} = 3.5 \times 10^{23} CCl_4 \text{ molecules}$

d) $19 \text{ g } C_6H_{12}O_6 \times \frac{1 \text{ mole } C_6H_{12}O_6}{180.18 \text{ g}} \times \frac{6.022 \times 10^{23} \text{molecules}}{1 \text{ mole } C_6H_{12}O_6}$

$= 6.4 \times 10^{22} \ C_6H_{12}O_6 \text{ molecules}$

55. $1.8 \times 10^{17} C_{12}H_{22}O_{11} \text{ molecules} \times \frac{1 \text{ mole } C_{12}H_{22}O_{11}}{6.022 \times 10^{23} \text{ molecules}} \times \frac{342.34 \text{ g}}{1 \text{ mole } C_{12}H_{22}O_{11}} \times$

$\frac{1000 \text{ mg}}{1 \text{ g}} = 0.10 \text{ mg } C_{12}H_{22}O_{11}$

57. mole → pennies → dollars → dollars/person

$$1 \text{ mol } \times \frac{6.022\times10^{23}\text{pennies}}{1 \text{ mol}} = 6.022\times10^{23}\text{pennies}$$

$$6.022\times10^{23}\text{pennies}\times\frac{1 \text{ dollar}}{100 \text{ pennies}} = 6.022\times10^{21}\text{dollars}$$

$$\frac{6.022\times10^{21}\text{dollars}}{6.6\times10^{9}\text{people}} = 9.1\times10^{11}\text{dollars/person}$$

$=9.1\times10^{2}$ billion dollars per person, each person would be a billionaire

Chemical Formulas as Conversion Factors

59. moles $CaCl_2$ → moles Cl

$$2.7 \text{ mol } CaCl_2 \times\frac{2 \text{ mol Cl}}{1 \text{ mol } CaCl_2} = 5.4 \text{ mol Cl}$$

61. a) $2.3 \text{ moles } H_2O\times\frac{1 \text{ mole O}}{1 \text{ mole } H_2O}=2.3 \text{ moles O}$

b) $1.2 \text{ moles } H_2O_2\times\frac{2 \text{ mole O}}{1 \text{ mole } H_2O_2} = 2.4 \text{ moles O}$

c) $0.9 \text{ moles } NaNO_3\times\frac{3 \text{ mole O}}{1 \text{ mole } NaNO_3}=2.7 \text{ moles O}$

d) $0.5 \text{ moles } Ca(NO_3)_2\times\frac{6 \text{ mole O}}{1 \text{ mole } Ca(NO_3)_2}=3.0 \text{ moles O}$

The correct answer is (d).

63. a) $2.5 \text{ mol } CH_4\times\frac{1 \text{ mole C}}{1 \text{ mole } CH_4} = 2.5 \text{ mol C}$

b) $0.115 \text{ mol } C_2H_6\times\frac{2 \text{ mole C}}{1 \text{ mole } C_2H_6} = 0.230 \text{ mol C}$

c) $5.67 \text{ mol } C_4H_{10}\times\frac{4 \text{ mole C}}{1 \text{ mole } C_4H_{10}} = 22.7 \text{ mol C}$

d) $25.1 \text{ mol } C_8H_{18}\times\frac{8 \text{ mole C}}{1 \text{ mole } C_8H_{18}} = 201 \text{ mol C}$

65. a) 2 moles H per mole of molecules; 8 H atoms present
b) 4 moles H per mole of molecules; 20 H atoms present
c) 3 moles H per mole of molecules; 9 H atoms present

67. a) $38.0 \text{ g } CF_2Cl_2 \times \frac{1 \text{ mol } CF_2Cl_2}{120.91 \text{ g}} \times \frac{2 \text{ mol Cl}}{1 \text{ mol } CF_2Cl_2} \times \frac{35.45 \text{ g}}{1 \text{ mol Cl}} = 22.3 \text{ g Cl}$

b) $38.0 \text{ g } CFCl_3 \times \frac{1 \text{ mol } CFCl_3}{137.36 \text{ g}} \times \frac{3 \text{ mol Cl}}{1 \text{ mol } CFCl_3} \times \frac{35.45 \text{ g}}{1 \text{ mol Cl}} = 29.4 \text{ g Cl}$

c) $38.0 \text{ g } C_2F_3Cl_3 \times \frac{1 \text{ mol } C_2F_3Cl_3}{187.37 \text{ g}} \times \frac{3 \text{ mol Cl}}{1 \text{ mol } C_2F_3Cl_3} \times \frac{35.45 \text{ g}}{1 \text{ mol Cl}} = 21.6 \text{ g Cl}$

d) $38.0 \text{ g } CF_3Cl \times \frac{1 \text{ mol } CF_3Cl}{104.46 \text{ g}} \times \frac{1 \text{ mol Cl}}{1 \text{ mol } CF_3Cl} \times \frac{35.45 \text{ g}}{1 \text{ mol Cl}} = 12.9 \text{ g Cl}$

69. a) $1.0 \times 10^3 \text{kg Fe} \times \frac{1000 \text{ g}}{1 \text{ kg}} \times \frac{1 \text{ mol Fe}}{55.85 \text{ g}} \times \frac{1 \text{ mol } Fe_2O_3}{2 \text{ mol Fe}} \times \frac{159.70 \text{ g}}{1 \text{ mol } Fe_2O_3} \times \frac{1 \text{ kg}}{1 \times 10^3 \text{ g}}$

$= 1.4 \times 10^3 \text{ kg } Fe_2O_3$

b) $1.0 \times 10^3 \text{kg Fe} \times \frac{1 \times 10^3 \text{g}}{1 \text{ kg}} \times \frac{1 \text{ mol Fe}}{55.85 \text{ g}} \times \frac{1 \text{ mol } Fe_3O_4}{3 \text{ mol Fe}} \times \frac{231.55 \text{ g}}{1 \text{ mol } Fe_3O_4} \times \frac{1 \text{ kg}}{1 \times 10^3 \text{g}}$

$= 1.4 \times 10^3 \text{ kg } Fe_3O_4$

c) $1.0 \times 10^3 \text{kg Fe} \times \frac{1 \times 10^3 \text{g}}{1 \text{ kg}} \times \frac{1 \text{ mol Fe}}{55.85 \text{ g}} \times \frac{1 \text{ mol } FeCO_3}{1 \text{ mol Fe}} \times \frac{115.86 \text{ g}}{1 \text{ mol } FeCO_3} \times \frac{1 \text{ kg}}{1 \times 10^3 \text{g}}$

$= 2.1 \times 10^3 \text{ kg } FeCO_3$

Mass Percent Composition

71. $\text{mass \% Sr} = \frac{2.45 \text{ g Sr}}{2.89 \text{ g SrO}} \times 100\% = 84.8\% \text{ Sr}$

73. $\text{mass \% Ca} = \frac{0.690 \text{ g Ca}}{1.912 \text{ g } CaCl_2} \times 100\% = 36.1\% \text{ Ca}$

$\text{mass \% Cl} = \frac{1.222 \text{ g Cl}}{1.912 \text{ g } CaCl_2} \times 100\% = 63.91\% \text{ Cl}$

75. $\frac{\text{mass F}}{28.5\text{g } CuF_2} \times 100\% = 37.42\% \text{ F} \Rightarrow \text{mass F} = \frac{(37.42\% \text{ F})(28.5 \text{ g } CuF_2)}{100\%} = 10.7 \text{ g}$

77. $\frac{3.0 \text{ mg F}}{\text{mass NaF}} \times 100\% = 45.24\% \text{ F} \Rightarrow \text{mass NaF} = \frac{(3.0 \text{ mg F})(100\%)}{45.24\% \text{ F}} = 6.6 \text{ mg NaF}$

Mass Percent Composition from Chemical Formula

79. Assume one mole of each compound and determine the mass % using molar masses.

a) $\text{mass \% N} = \dfrac{28.02\text{ g N}}{44.02\text{ g N}_2\text{O}} \times 100\% = 63.65\%\text{ N}$

b) $\text{mass \% N} = \dfrac{14.01\text{ g N}}{30.01\text{ g NO}} \times 100\% = 46.68\%\text{ N}$

c) $\text{mass \% N} = \dfrac{14.01\text{ g N}}{46.01\text{ g NO}_2} \times 100\% = 30.45\%\text{ N}$

d) $\text{mass \% N} = \dfrac{28.02\text{ g N}}{108.02\text{ g N}_2\text{O}_5} \times 100\% = 25.94\%\text{ N}$

81. Assume 1 mole of each compound, base mass % molar masses.

a) $\text{mass \% C} = \dfrac{24.02\text{ g C}}{60.06\text{ g C}_2\text{H}_4\text{O}_2} \times 100\% = 39.99\%\text{ C}$

$\text{mass \% H} = \dfrac{4.04\text{ g H}}{60.06\text{ g C}_2\text{H}_4\text{O}_2} \times 100\% = 6.73\%\text{ H}$

$\text{mass \% O} = \dfrac{32.00\text{ g O}}{60.06\text{ g C}_2\text{H}_4\text{O}_2} \times 100\% = 53.28\%\text{ O}$

b) $\text{mass \% C} = \dfrac{12.01\text{ g C}}{46.03\text{ g CH}_2\text{O}_2} \times 100\% = 26.09\%\text{ C}$

$\text{mass \% H} = \dfrac{2.02\text{ g H}}{46.03\text{ g CH}_2\text{O}_2} \times 100\% = 4.39\%\text{ H}$

$\text{mass \% O} = \dfrac{32.00\text{ g O}}{46.03\text{ g CH}_2\text{O}_2} \times 100\% = 69.52\%\text{ O}$

c) $\text{mass \% C} = \dfrac{36.03\text{ g C}}{59.13\text{ g C}_3\text{H}_9\text{N}} \times 100\% = 60.93\%\text{ C}$

$\text{mass \% H} = \dfrac{9.09\text{ g H}}{59.13\text{ g C}_3\text{H}_9\text{N}} \times 100\% = 15.4\%\text{ H}$

$\text{mass \% N} = \dfrac{14.01\text{ g N}}{59.13\text{ g C}_3\text{H}_9\text{N}} \times 100\% = 23.69\%\text{ N}$

$$\text{d) mass \% C} = \frac{48.04 \text{ g C}}{88.18 \text{ g } C_4H_{12}N_2} \times 100\% = 54.48\% \text{ C}$$

$$\text{mass \% H} = \frac{12.12 \text{ g H}}{88.18 \text{ g } C_4H_{12}N_2} \times 100\% = 13.74\% \text{ H}$$

$$\text{mass \% N} = \frac{28.02 \text{ g N}}{88.18 \text{ g } C_4H_{12}N_2} \times 100\% = 31.78\% \text{ N}$$

83. $$\text{mass \% Fe} = \frac{111.7 \text{ g Fe}}{159.70 \text{ g } Fe_2O_3} \times 100\% = 69.94\% \text{ Fe}$$

$$\text{mass \% Fe} = \frac{167.55 \text{ g Fe}}{231.55 \text{ g } Fe_3O_4} \times 100\% = 72.36\% \text{ Fe}$$

$$\text{mass \% Fe} = \frac{55.85 \text{ g Fe}}{115.86 \text{ g } FeCO_3} \times 100\% = 48.20\% \text{ Fe}$$

The magnetite ore has the highest iron content.

Calculating Empirical Formulas

85. $$1.78 \text{ g N} \times \frac{1 \text{ mole N}}{14.01 \text{ g}} = 0.127 \text{ mol N}$$

$$4.05 \text{ g O} \times \frac{1 \text{ mole O}}{16.00 \text{ g}} = 0.253 \text{ mol O}$$

$$N_{\frac{0.127}{0.127}}O_{\frac{0.253}{0.127}} = NO_2$$

87. a) $$1.245 \text{ g Ni} \times \frac{1 \text{ mole Ni}}{58.69 \text{ g}} = 0.02121 \text{ mol Ni}$$

$$5.381 \text{ g I} \times \frac{1 \text{ mole I}}{126.90 \text{ g}} = 0.04240 \text{ mol I}$$

$$Ni_{\frac{0.02121}{0.02121}}I_{\frac{0.04240}{0.02121}} = NiI_2$$

b) $$1.443 \text{ g Se} \times \frac{1 \text{ mole Se}}{78.96 \text{ g}} = 0.01828 \text{ mol Se}$$

$$5.841 \text{ g Br} \times \frac{1 \text{ mole Br}}{79.90 \text{ g}} = 0.07310 \text{ mol Br}$$

$$Se_{\frac{0.01828}{0.01828}}Br_{\frac{0.07310}{0.01828}} = SeBr_4$$

c) $2.128 \text{ g Be} \times \frac{1 \text{ mole Be}}{9.01 \text{ g}} = 0.236 \text{ mol Be}$

$7.557 \text{ g S} \times \frac{1 \text{ mole S}}{32.07 \text{ g}} = 0.2356 \text{ mol S}$

$15.107 \text{ g O} \times \frac{1 \text{ mole O}}{16.00 \text{ g}} = 0.9442 \text{ mol O}$

$Be_{\frac{0.236}{0.2356}} S_{\frac{0.2356}{0.2356}} O_{\frac{0.9442}{0.2356}} = BeSO_4$

89. Assume a 100 gram sample, percentage composition is then equal to the number of grams of each element.

$54.50 \text{ g C} \times \frac{1 \text{ mole C}}{12.01 \text{ g}} = 4.538 \text{ mol C}$

$13.73 \text{ g H} \times \frac{1 \text{ mole H}}{1.01 \text{ g}} = 13.6 \text{ mol H}$

$31.77 \text{ g N} \times \frac{1 \text{ mole N}}{14.01 \text{ g}} = 2.268 \text{ mol N}$

$C_{\frac{4.538}{2.268}} H_{\frac{13.6}{2.268}} N_{\frac{2.268}{2.268}} \Rightarrow C_2H_6N$

91. Assume a 100 gram sample, percentage composition is then equal to the number of grams of each element.

a) $62.04 \text{ g C} \times \frac{1 \text{ mole C}}{12.01 \text{ g}} = 5.166 \text{ moles C}$

$10.41 \text{ g H} \times \frac{1 \text{ mole g H}}{1.01 \text{ g H}} = 10.3 \text{ moles H}$

$27.55\text{g O} \times \frac{1 \text{ mole g O}}{16.00 \text{ g O}} = 1.722 \text{ moles O}$

$C_{\frac{5.166}{1.722}} H_{\frac{10.3}{1.722}} O_{\frac{1.722}{1.722}} = C_3H_6O$

b) $58.80 \text{ g C} \times \frac{1 \text{ mole C}}{12.01 \text{ g}} = 4.896 \text{ moles C}$

$9.87 \text{ g H} \times \frac{1 \text{ mole g H}}{1.01 \text{ g H}} = 9.77 \text{ moles H}$

$31.33\text{g O} \times \frac{1 \text{ mole g O}}{16.00 \text{ g O}} = 1.958 \text{ moles O}$

$C_{\frac{4.896}{1.958}} H_{\frac{9.77}{1.958}} O_{\frac{1.958}{1.958}} = C_{2.5}H_5O_1$; Empirical Formula$=2 \times (C_{2.5}H_5O_1) = C_5H_{10}O_2$

c) $71.98 \text{ g C} \times \dfrac{1 \text{ mole C}}{12.01 \text{ g}} = 5.993 \text{ moles C}$

$6.71 \text{ g H} \times \dfrac{1 \text{ mole g H}}{1.01 \text{ g H}} = 6.64 \text{ moles H}$

$21.31 \text{g O} \times \dfrac{1 \text{ mole g O}}{16.00 \text{ g O}} = 1.332 \text{ moles O}$

$C_{\frac{5.993}{1.332}} H_{\frac{6.64}{1.332}} O_{\frac{1.332}{1.332}} = C_{4.50}H_5O_1$; Empirical Formula$=2 \times (C_{4.50}H_5O_1) = C_9H_{10}O_2$

93. $1.45 \text{ g P} \times \dfrac{1 \text{ mole P}}{30.97 \text{ g P}} = 0.0468 \text{ mole P}$

mass $O = 2.57 - 1.45 = 1.12 \text{ g O}$

$1.12 \text{ g O} \times \dfrac{1 \text{ mole O}}{16.00 \text{ g O}} = 0.0700 \text{ mole O}$

$P_{\frac{0.0468}{0.0468}} O_{\frac{0.0700}{0.0468}} = P_1O_{1.50}$; Empirical Formula $= 2 \times (P_1O_{1.50}) = P_2O_3$

95. $0.77 \text{ mg N} \times \dfrac{1 \text{ g}}{1000 \text{ mg}} \times \dfrac{1 \text{ mole N}}{14.01 \text{ g N}} = 5.5 \times 10^{-5} \text{ mole N}$

mass $Cl = 6.61 - 0.77 = 5.84 \text{ mg Cl}$

$5.84 \text{ mg Cl} \times \dfrac{1 \text{ g}}{1000 \text{ mg}} \times \dfrac{1 \text{ mole Cl}}{35.45 \text{ g Cl}} = 1.65 \times 10^{-4} \text{ mole Cl}$

$N_{\frac{5.50 \times 10^{-5}}{5.50 \times 10^{-5}}} Cl_{\frac{1.65 \times 10^{-4}}{5.50 \times 10^{-5}}} = NCl_3$

Calculating Molecular Formulas

97. $\dfrac{\text{Molar Mass}}{\text{Empirical Mass}} = \text{Multiplier} \Rightarrow \dfrac{56.11}{14.03} = 4 \Rightarrow 4 \times (CH_2) = C_4H_8$

99. $\dfrac{\text{Molar Mass}}{\text{Empirical Mass}} = \text{Multiplier}$

a) $\dfrac{284.77}{47.46} = 6 \Rightarrow 6 \times (CCl) = C_6Cl_6$

b) $\dfrac{131.39}{131.38} = 1 \Rightarrow 1 \times (C_2HCl_3) = C_2HCl_3$

c) $\dfrac{181.44}{60.48} = 3 \Rightarrow 3 \times (C_2HCl) = C_6H_3Cl_3$

Cumulative Problems

101. Volume=$L^3 \Rightarrow (1.42 \text{ cm})^3 = 2.86 \text{ cm}^3$

mass=density × volume $\Rightarrow (8.96 \text{ g Cu/cm}^3)(2.86 \text{ cm}^3) = 25.6 \text{ g Cu}$

$$25.6 \text{ g Cu} \times \frac{1 \text{ mole Cu}}{63.55 \text{ g}} \times \frac{6.022 \times 10^{23} \text{ Cu atoms}}{1 \text{ mole Cu}} = 2.43 \times 10^{23} \text{ Cu atoms}$$

103. Volume: $1\text{ml}=1\text{cm}^3 \Rightarrow 0.05 \text{ cm}^3 \ H_2O$

mass = density × volume $\Rightarrow (1.0 \text{g } H_2O/\text{cm}^3)(0.05 \text{ cm}^3) = 0.05 \text{ g } H_2O$

$$0.05 \text{ g } H_2O \times \frac{1 \text{ mole } H_2O}{18.02 \text{ g}} \times \frac{6.022 \times 10^{23} \text{particles}}{1 \text{ mole } H_2O} = 2 \times 10^{21} H_2O \text{ molecules}$$

105.

Substance	Mass	Moles	Number of Particles
Ar	0.018 g	4.5×10^{-4}	2.7×10^{20}
NO_2	8.33×10^{-3} g	1.81×10^{-4}	1.09×10^{20}
K	22.4 mg	5.73×10^{-4}	3.45×10^{20}
C_8H_{18}	3.76 kg	32.9	1.98×10^{25}

107. a) CuI_2; Formula Mass = 317.35

$$\text{Mass \% Cu} = \frac{63.55}{317.35} \times 100\% = 20.03\% \text{ Cu}$$

$$\text{Mass \% I} = \frac{253.8}{317.35} \times 100\% = 79.97\% \text{ I}$$

b) $NaNO_3$; Formula Mass = 85.00

$$\text{Mass \% Na} = \frac{22.99}{85.00} \times 100\% = 27.05\%$$

$$\text{Mass \% N} = \frac{14.01}{85.00} \times 100\% = 16.48\%$$

$$\text{Mass \% O} = \frac{48.00}{85.00} \times 100\% = 56.47\% \text{ O}$$

c) $PbSO_4$; Formula Mass = 303.3

$$\text{Mass \% Pb} = \frac{207.2}{303.3} \times 100\% = 68.32\% \text{ Pb}$$

$$\text{Mass \% S} = \frac{32.07}{303.3} \times 100\% = 10.57\% \text{ S}$$

$$\text{Mass \% O} = \frac{64.00}{303.3} \times 100\% = 21.10\% \text{ O}$$

d) CaF_2; Formula Mass = 78.08

$$\text{Mass \% Ca} = \frac{40.08}{78.08} \times 100\% = 51.33\% \text{ Ca}$$

$$\text{Mass \% F} = \frac{38.00}{78.08} \times 100\% = 48.67\% \text{ F}$$

109. Step 1: Determine how much Fe_2O_3 would be needed to obtain $1x10^3$ kg of iron.

$$1.0x10^3 \text{kg Fe} \times \frac{1000\text{g}}{1 \text{ kg}} \times \frac{1 \text{ mol Fe}}{55.85 \text{ g}} \times \frac{1 \text{ mol } Fe_2O_3}{2 \text{ mol Fe}} \times \frac{159.70 \text{ g}}{1 \text{ mol } Fe_2O_3} \times \frac{1 \text{ kg}}{1000\text{g}}$$

$= 1.4 \times 10^3$ kg Fe_2O_3

Step 2: Based on the ore being 78% Fe_2O_3, determine the amount of rock needed for processing. Remember 78% Fe_2O_3 can be used as a conversion factor because 78 kg Fe_2O_3 is obtained for every 100 kg of ore mined.

$$1.4 \times 10^3 \text{ kg } Fe_2O_3 \times \frac{100 \text{ kg rock}}{78 \text{ kg } Fe_2O_3} = 1.8 \times 10^3 \text{kg rock}$$

111. $$\frac{12 \text{ kg } CHF_2Cl}{1 \text{ mo}} \times 12 \text{ mo} \times \frac{1000 \text{ g}}{1 \text{ kg}} \times \frac{1 \text{ mol } CHF_2Cl}{86.47 \text{ g}} \times \frac{1 \text{ mol Cl}}{1 \text{ mol } CHF_2Cl} \times \frac{1 \text{ kg}}{1000 \text{ g}}$$

= 59 kg Cl

113. $$1.0 \text{ L } H_2O \times \frac{1000 \text{ cm}^3}{1 \text{ L}} \times \frac{1 \text{ g } H_2O}{1 \text{ cm}^3} \times \frac{1 \text{ mol } H_2O}{18.0 \text{ g}} \times \frac{2 \text{ mol H}}{1 \text{ mol } H_2O} \times \frac{1.008 \text{ g}}{1 \text{ mol H}} = 1.1 \times 10^2 \text{ g H}$$

115.

Formula	Molar Mass	%C (by mass)	%H (by mass)
C_2H_4	28.06	85.60%	14.40%
C_4H_{10}	58.12	82.66%	17.34%
C_4H_8	56.12	85.60%	14.40%
C_3H_8	44.11	81.71%	18.29%

117. Assume a 100 gram sample, % composition then equals the number of grams of each element.

$$55.80 \text{ g C} \times \frac{1 \text{ mole C}}{12.01 \text{ g}} = 4.646 \text{ moles C}$$

$$7.03 \text{ g H} \times \frac{1 \text{ mole H}}{1.01 \text{ g H}} = 6.96 \text{ moles H}$$

$$37.17\text{g O} \times \frac{1 \text{ mole O}}{16.00 \text{ g O}} = 2.323 \text{ moles O}$$

$$C_{\frac{4.646}{2.323}} H_{\frac{6.96}{2.323}} O_{\frac{2.323}{2.323}} = C_2H_3O$$

$$\frac{\text{Molar Mass}}{\text{Empirical Mass}} = \text{Multiplier} \Rightarrow \frac{86.09}{43.04} = 2$$

$$\text{Molecular Formula}=2 \times (C_2H_3O) = C_4H_6O_2$$

119. Assume a 100 gram sample, % composition then equals the number of grams of each element.

$$74.03 \text{ g C} \times \frac{1 \text{ mole C}}{12.01 \text{ g}} = 6.164 \text{ mole C}$$

$$8.70 \text{ g H} \times \frac{1 \text{ mole H}}{1.01 \text{ g H}} = 8.61 \text{ moles H}$$

$$17.27 \text{ g N} \times \frac{1 \text{ mole N}}{14.01 \text{ g N}} = 1.233 \text{ moles N}$$

$$C_{\frac{6.164}{1.233}} H_{\frac{8.630}{1.233}} N_{\frac{1.233}{1.233}} = C_5H_7N$$

$$\frac{\text{Molar Mass}}{\text{Empirical Mass}} = \text{Multiplier} \Rightarrow \frac{162.23}{81.12} = 2$$

$$\text{Molecular Formula}=2 \times (C_5H_7N) = C_{10}H_{14}N_2$$

121. The mass of the sample consists of KBr and KI as shown in equation 1:

Eqn 1: Mass KBr + Mass KI = 5.00 g

The mass of KBr and KI can be calculated by multiplying the moles of each compound by the formula mass of that compound as shown in equation 2.

Eqn 2: (moles KBr)(FM KBr) + (moles KI)(FM KI) = 5.00g

The sample contains 1.51 g K which corresponds to 0.0386193 mol K (note: to prevent introducing errors into the calculation, numbers will not be rounded to account for significant figures until the final answer). Because the sample consists of KBr and KI which have 1 mole of K per mole of the compound, the following can be written:

Eqn 3: moles KBr + moles KI = 0.0386193

Using equations 2 and 3 we have the situation of two equations and two unknowns to solve.

(0.0386193 - mol KI)(119.00) + moles KI(166.00)=5.00
4.5956967 - 119.00 mol KI + 166.00 mol KI = 5.00
47.00 mol KI = 0.4043033
mol KI = 0.4043033/47.00
mol KI = 0.008602198
moles KBr + 0.008602198 = 0.0386193
mol KBr= 0.03001710

%KI = (0.008602198 mol KI × 166 g/mol KI)/5.00 ×100% = 28.6% KI
%KBr = (0.03001710 mol KBr × 119 g/mol KBr)/5.00 ×100%=71.4% KBr

123. $g\ C_2H_6S \rightarrow mol\ C_2H_6S \rightarrow mol\ SO_2 \rightarrow g\ SO_2$

Reaction: $2C_2H_6S + 9O_2 \rightarrow 4CO_2 + 6H_2O + 2SO_2$

$$28.7 g C_2H_6S \times \frac{1\text{ mol } C_2H_6S}{62.15\text{ g } C_2H_6S} \times \frac{2\text{ mol } SO_2}{2\text{ mol } C_2H_6S} \times \frac{64.07\text{ g } SO_2}{1\text{ mol } SO_2} = 29.6\text{ g } SO_2$$

125. mass ore →mass Fe_2O_3 →mass Fe
10.0 kg Ore × 0.38 = 3.8 kg Fe_2O_3
Fe_2O_3 = 111.7 g Fe/159.7 g Fe_2O_3 = 69.94% Fe
3.8 kg Fe_2O_3 × 0.6994 = 2.7 kg Fe

127. $V=4/3\pi(7\times10^{8}\text{m})^3 = 1.4\times10^{27}\text{m}^3$; convert to $\text{cm}^3 \Rightarrow$

$$1.4\times10^{27}\text{m}^3\times\frac{(100\text{ cm})^3}{(1\text{ m})^3} = 1.4\times10^{33}\text{cm}^3\text{; calculate grams of H} \Rightarrow$$

$$1.4\times10^{33}\text{cm}^3\text{ H}\times\frac{1.4\text{ g H}}{1\text{ cm}^3} = 2.0\times10^{33}\text{g H; convert to moles \& atoms} \Rightarrow$$

$$2.0\times10^{33}\text{g H}\times\frac{1\text{ mole H}}{1.008\text{ g}}\times\frac{6.022\times10^{23}\text{atoms}}{1\text{ mole H}} = 1\times10^{57}\text{H atoms per star}$$

b) $$\frac{1\times10^{57}\text{H atoms}}{1\text{ star}}\times\frac{1\times10^{11}\text{stars}}{\text{galaxy}} = 1\times10^{68}\text{ H atoms per galaxy}$$

c) $$\frac{1\times10^{68}\text{H atoms}}{1\text{ galaxy}}\times\frac{1\times10^{11}\text{galaxies}}{\text{universe}} = 1\times10^{79}\text{ H atoms in the universe}$$

129. Assume a 100 gram sample, % composition is then equal to number of grams of each element.

$$95.02\text{ g C}\times\frac{1\text{ mole C}}{12.01\text{ g}} = 7.912\text{ mole C}$$

$$4.98\text{ g H}\times\frac{1\text{ mole H}}{1.01\text{ g H}} = 4.93\text{ moles H}$$

$$C_{\frac{7.912}{4.93}}H_{\frac{4.93}{4.93}} = C_{1.6}H_1\text{; Empirical formula}=5\times(C_{1.6}H_1) = C_8H_5$$

$$\frac{\text{Molar Mass}}{\text{Empirical Mass}} = \text{Multiplier} \Rightarrow \frac{202.23}{101.12} = 2$$

Molecular Formula $=2\times(C_8H_5)=C_{16}H_{10}$

Chemical Reactions

7

Questions

1. In a reaction, one or more substances change into different substances. Many examples can be given, such as the neutralization of acid with a base (vinegar + baking soda).

3. The following constitute the main evidence that a chemical reaction has occurred:
   1) a color change
   2) formation of a solid
   3) formation of a gas
   4) emission of light
   5) emission of *(or absorption of)* heat

5. a) gas
   b) liquid
   c) solid
   d) aqueous (dissolved in water)

7.

	Element	Reactants	Products
a)	Ag	4	4
	O	2	2
	C	1	1
	Yes, the equation is balanced.		
b)	Pb	1	1
	N	2	2
	O	6	6
	Na	2	2
	Cl	2	2
	Yes, the equation is balanced.		
c)	C	3	3
	H	8	8
	O	2	10
	No, the equation is not balanced.		

9. A soluble compound will dissolve in solution in appreciable quantities, while an insoluble compound will hardly dissolve in solution at all.

11. Polyatomic ions dissolve in water and dissociate into the ions that make up the compound. The polyatomic ion group stays together as one particle. For example, $NaNO_3$ in water will form Na^+*(aq)* and NO_3^-*(aq)*.

13. The solubility rules are as follows:

Soluble Compounds

1. Any compound containing Li^+, Na^+, K^+, or NH_4^+.
2. Any compound containing NO_3^- or $C_2H_3O_2^-$.
3. Most compounds containing Cl^- Br^-, or I^-.
   Except with Ag^+, Hg_2^{2+}, or Pb^{2+}, they form insoluble compounds.
4. Most compounds with SO_4^{2-}
   Except with Sr^{2+}, Ba^{2+}, Pb^{2+}, or Ca^{2+}, they form insoluble compounds.

Insoluble Compounds

1. Most compounds containing OH^- or S^{2-}
   Except with Li^+, Na^+, K^+, or NH_4^+, they form soluble compounds.
   Except when S^{2-} is with Sr^{2+}, Ba^{2+}, or Ca^{2+}, they form soluble compounds.
   Except when OH^- is with Sr^{2+}, Ba^{2+}, or Ca^{2+}, they form slightly soluble compounds.

2. Most compounds containing CO_3^{2-} or PO_4^{3-}
   Except with Li^+, Na^+, K^+, or NH_4^+, they form soluble compounds.

These rules are useful because there is no way to predict solubility from looking at the periodic table. These rules allow us to know what compounds will dissolve in water and which ones will not, which is critical when writing chemical equations.

15. The precipitate in a precipitation reaction will always be the insoluble compound. Otherwise, it would dissolve in water, no solid would form and there would be no precipitation reaction.

17. An acid-base reaction involves a neutralization of H^+ *(aq)* ions from an acid and OH^- *(aq)* ions from a base to form water. The counter ions form a salt. Consider the following reaction: $NaOH\ (aq) + HCl\ (aq) \rightarrow H_2O\ (l)$ and $NaCl\ (aq)$. The base is NaOH, the acid is HCl.

19. A gas evolution reaction occurs when one of the products of a chemical reaction is a gas. An example is: $2HCl(aq) + CaS(aq) \rightarrow H_2S(g) + CaCl_2(aq)$

21. A combustion reaction involves the reaction of a substance with O_2 to form oxygen-containing compounds, often including water.
An example is: $2C_2H_6(g) + 7O_2(g) \rightarrow 4CO_2(g) + 6H_2O(g)$

23. In a synthesis reaction simpler substances combine to form more complex substances, for example: $2Na(s) + Cl_2(g) \rightarrow 2NaCl(s)$. A decomposition reaction occurs when a complex substance decomposes to form simpler substances, for example: $2H_2O(l) \rightarrow 2H_2(g) + O_2(g)$.

Problems

Evidence of Chemical Reactions

25. a) A chemical reaction because the initial compounds change to form a solid and a color change occurs.
    b) Not a chemical reaction because the initial compound did not change into another substance.
    c) A chemical reaction because the initial compounds change to form a solid.
    d) A chemical reaction because the initial compounds change to form a gas and other new compounds.

27. Yes, a chemical reaction has occurred because the bubbles that formed are a new compound formed as a product of the reaction.

29. Yes, a chemical reaction has occurred because the color change in the hair is due to forming new compounds in the hair itself.

Writing and Balancing Chemical Equations

31. By adding a subscript, the nature of the chemical is changed and the equation no longer accurately represents the chemical reaction it is supposed to describe. The proper method of balancing a reaction is to add coefficients in front of each reactant or product compound. The balanced reaction is: $2H_2O\ (l) \rightarrow 2H_2(g) + O_2(g)$.

33. a) $PbS(s) + 2HCl(aq) \rightarrow PbCl_2(s) + H_2S(g)$
    b) $CO(g) + 3H_2(g) \rightarrow CH_4(g) + H_2O(l)$
    c) $Fe_2O_3(s) + 3H_2(g) \rightarrow 2Fe(s) + 3H_2O(l)$
    d) $4NH_3(g) + 5O_2(g) \rightarrow 4NO(g) + 6H_2O(g)$

35. a) $Mg(s) + 2CuNO_3(aq) \rightarrow Mg(NO_3)_2(aq) + 2Cu(s)$
    b) $2N_2O_5(g) \rightarrow 4NO_2(g) + O_2(g)$
    c) $Ca(s) + 2HNO_3(aq) \rightarrow Ca(NO_3)_2(aq) + H_2(g)$
    d) $2CH_3OH(l) + 3O_2(g) \rightarrow 2CO_2(g) + 4H_2O(g)$

37. $2H_2(g) + O_2(g) \rightarrow 2H_2O(l)$; $Cl(g) + O_3(g) \rightarrow ClO(g) + O_2(g)$

39. $2Na(s) + 2H_2O(l) \rightarrow H_2(g) + 2NaOH(aq)$

41. $2SO_2(g) + O_2(g) + 2H_2O(l) \rightarrow 2H_2SO_4(aq)$

43. $V_2O_5(s) + 2H_2(g) \rightarrow V_2O_3(s) + 2H_2O(l)$

45. $C_{12}H_{22}O_{11}(aq) + H_2O(l) \rightarrow 4C_2H_5OH(aq) + 4CO_2(g)$

47. a) $Na_2S(aq) + Cu(NO_3)_2(aq) \rightarrow 2NaNO_3(aq) + CuS(s)$
b) $4HCl(aq) + O_2(g) \rightarrow 2H_2O(l) + 2Cl_2(g)$
c) $2H_2(g) + O_2(g) \rightarrow 2H_2O(l)$
d) $FeS(s) + 2HCl(aq) \rightarrow FeCl_2(aq) + H_2S(g)$

49. a) $BaO_2(s) + H_2SO_4(aq) \rightarrow BaSO_4(s) + H_2O_2(aq)$
b) $2Co(NO_3)_3(aq) + 3(NH_4)_2S(aq) \rightarrow Co_2S_3(s) + 6NH_4NO_3(aq)$
c) $Li_2O(s) + H_2O(l) \rightarrow 2LiOH(aq)$
d) $Hg_2(C_2H_3O_2)_2(aq) + 2KCl(aq) \rightarrow Hg_2Cl_2(s) + 2KC_2H_3O_2(aq)$

51. a) $2Rb(s) + 2H_2O(l) \rightarrow 2RbOH(aq) + H_2(g)$
b) Ok
c) $2NiS(s) + 3O_2(g) \rightarrow 2NiO(s) + 2SO_2(g)$
d) $3PbO(s) + 2NH_3(g) \rightarrow 3Pb(s) + N_2(g) + 3H_2O(l)$

53. $C_6H_{12}O_6(aq) + 6O_2(g) \rightarrow 6CO_2(g) + 6H_2O(l)$

55. $2NO(g) + 2CO(g) \rightarrow N_2(g) + 2CO_2(g)$

Solubility

57. a) Soluble: Na^+, $C_2H_3O_2^-$
b) Soluble: Sn^{2+}, NO_3^-
c) Insoluble
d) Soluble: Na^+, PO_4^{3-}

59. Ag^+ with Cl^-, $AgCl$
Ba^{2+} with SO_4^{2-}, $BaSO_4$
Cu^{2+} with CO_3^{2-}, $CuCO_3$
Fe^{3+} with S^{2-}, Fe_2S_3

61.

Soluble	Insoluble	Soluble	Insoluble
K_2S	Hg_2I_2	K_2SO_4	$PbSO_4$
BaS	$Cu_3(PO_4)_2$	SrS	$PbCl_2$
NH_4Cl	MgS	Li_2S	Hg_2Cl_2
Na_2CO_3	$CaSO_4$		

Precipitation Reactions

63. a) no reaction
    b) $K_2SO_4(aq) + BaBr_2(aq) \rightarrow BaSO_4(s) + 2\ KBr(aq)$
    c) $2NaCl(aq) + Hg_2(C_2H_3O_2)_2(aq) \rightarrow 2NaC_2H_3O_2(aq) + Hg_2Cl_2(s)$
    d) no reaction

65. a) $Na_2CO_3(aq) + Pb(NO_3)_2(aq) \rightarrow PbCO_3(s) + 2NaNO_3(aq)$
    b) $K_2SO_4(aq) + Pb(C_2H_3O_2)_2(aq) \rightarrow PbSO_4(s) + 2KC_2H_3O_2(aq)$
    c) $Cu(NO_3)_2(aq) + BaS(aq) \rightarrow CuS(s) + Ba(NO_3)_2(aq)$
    d) no reaction

67. a) correct
    b) no reaction
    c) correct
    d) $Pb(NO_3)_2(aq) + 2\ LiCl(aq) \rightarrow 2\ LiNO_3(aq) + PbCl_2(s)$

Ionic and Net Ionic Equations

69. Spectator Ions: NO_3^-, K^+

71. a) Complete Ionic:
    $Ag^+(aq) + NO_3^-(aq) + K^+(aq) + Cl^-(aq) \rightarrow AgCl(s) + K^+(aq) + NO_3^-(aq)$
    Net Ionic: $Ag^+(aq) + Cl^-(aq) \rightarrow AgCl(s)$
    b) Complete Ionic:
    $Ca^{2+}(aq) + S^{2-}(aq) + Cu^{2+}(aq) + 2Cl^-(aq) \rightarrow CuS(s) + Ca^{2+}(aq) + 2Cl^-(aq)$
    Net Ionic: $S^{2-}(aq) + Cu^{2+}(aq) \rightarrow CuS(s)$
    c) Complete Ionic:
    $Na^+(aq) + OH^-(aq) + H^+(aq) + NO_3^-(aq) \rightarrow H_2O(l) + Na^+(aq) + NO_3^-(aq)$
    Net Ionic: $OH^-(aq) + H^+(aq) \rightarrow H_2O(l)$
    d) Complete Ionic:
    $6K^+(aq) + 2PO_4^{3-}(aq) + 3Ni^{2+}(aq) + 6Cl^-(aq) \rightarrow Ni_3(PO_4)_2(s) + 6K^+(aq) + 6Cl^-(aq)$
    Net Ionic: $2PO_4^{3-}(aq) + 3Ni^{2+}(aq) \rightarrow Ni_3(PO_4)_2(s)$

    Net Ionic: $NH_4^+(aq) + OH^-(aq) \rightarrow H_2O(l) + NH_3(g)$

73. Complete Ionic:
    $Hg_2^{2+}(aq) + 2NO_3^-(aq) + 2Na^+(aq) + 2Cl^-(aq) \rightarrow Hg_2Cl_2(s) + 2Na^+(aq) + NO_3^-(aq)$
    Net Ionic: $Hg_2^{2+}(aq) + 2Cl^-(aq) \rightarrow Hg_2Cl_2(s)$

75. a) Complete Ionic:

$2Na^+(aq) + CO_3^{2-}(aq) + Pb^{2+}(aq) + 2NO_3^-(aq) \rightarrow PbCO_3(s) + 2Na^+(aq) + 2NO_3^-(aq)$

Net Ionic: $CO_3^{2-}(aq) + Pb^{2+}(aq) \rightarrow PbCO_3(s)$

b) Complete Ionic:

$2K^+(aq) + SO_4^{2-}(aq) + Pb^{2+}(aq) + 2C_2H_3O_2(aq) \rightarrow PbSO_4(s) + 2K^+(aq) + 2C_2H_3O_2^-(aq)$

Net Ionic: $SO_4^{2-}(aq) + Pb^{2+}(aq) \rightarrow PbSO_4(s)$

c) Complete Ionic:

$Cu^{2+}(aq) + 2NO_3^-(aq) + Ba^{2+}(aq) + S^{2-}(aq) \rightarrow CuS(s) + Ba^{2+}(aq) + 2NO_3^-(aq)$

Net Ionic: $Cu^{2+}(aq) + S^{2-}(aq) \rightarrow CuS(s)$

d) no reaction

Acid-Base and Gas Evolution Reactions

77. Molecular: $HCl(aq) + KOH(aq) \rightarrow KCl(aq) + H_2O(l)$

Net Ionic: $H^+(aq) + OH^-(aq) \rightarrow H_2O(l)$

79. a) $2HCl(aq) + Ba(OH)_2(aq) \rightarrow 2H_2O(l) + BaCl_2(aq)$

b) $H_2SO_4(aq) + 2KOH(aq) \rightarrow 2H_2O(l) + K_2SO_4(aq)$

c) $HClO_4(aq) + NaOH(aq) \rightarrow H_2O(l) + NaClO_4(aq)$

81. a) $HBr(aq) + NaHCO_3(aq) \rightarrow CO_2(g) + H_2O(l) + NaBr(aq)$

b) $NH_4I(aq) + KOH(aq) \rightarrow H_2O(l) + NH_3(g) + KI(aq)$

c) $2HNO_3(aq) + K_2SO_3(aq) \rightarrow SO_2(g) + H_2O(l) + 2KNO_3(aq)$

d) $2HI(aq) + Li_2S(aq) \rightarrow H_2S(g) + 2LiI(aq)$

Oxidation-Reduction and Combustion

83. b) metal reacting with a nonmetal

d) transfer of electrons between reactants

85. a) $2C_2H_6(g) + 7O_2(g) \rightarrow 4CO_2(g) + 6H_2O(g)$

b) $2Ca(s) + O_2(g) \rightarrow 2CaO(s)$

c) $2C_3H_8O(l) + 9O_2(g) \rightarrow 6CO_2(g) + 8H_2O(g)$

d) $2C_4H_{10}S(l) + 15O_2(g) \rightarrow 8CO_2(g) + 10H_2O(g) + 2SO_2(g)$

87. a) $2Ag(s) + Br_2(g) \rightarrow 2AgBr(s)$

b) $2K(s) + Br_2(g) \rightarrow 2KBr(s)$

c) $2Al(s) + 3Br_2(g) \rightarrow 2AlBr_3(s)$

d) $Ca(s) + Br_2(g) \rightarrow CaBr_2(s)$

Classifying Chemical Reactions by What Atoms Do

89. a) double displacement
    b) synthesis
    c) single displacement
    d) decomposition

91. a) synthesis
    b) decomposition
    c) synthesis

Cumulative Problems

93. a) Complete Ionic:
    $2Na^+(aq) + 2I^-(aq) + Hg_2^{2+}(aq) + 2NO_3^-(aq) \rightarrow Hg_2I_2(s) + 2Na^+(aq)+$
    $2NO_3^-(aq)$
    Net Ionic:
    $2I^-(aq) + Hg_2^{2+}(aq) \rightarrow Hg_2I_2(s)$
    b) Complete Ionic:
    $2H^+(aq)+ 2ClO_4^-(aq)+ Ba^{2+}(aq)+ 2OH^-(aq)\rightarrow 2H_2O(l)+ 2ClO_4^-(aq)+ Ba^{2+}(aq)$
    Net Ionic:
    $H^+(aq) + OH^-(aq) \rightarrow H_2O(l)$
    c) no reaction
    d) Complete Ionic:
    $2H^+(aq)+ 2Cl^-(aq)+ 2Li^+(aq)+ CO_3^{2-}(aq)\rightarrow CO_2(g)+ H_2O(l)+ 2Li^+(aq)+ 2Cl^-(aq)$
    Net Ionic:
    $2H^+(aq) + CO_3^{2-}(aq) \rightarrow CO_2(g) + H_2O(l)$

95. a) No reaction
    b) No reaction
    c) Complete Ionic:
    $H^+(aq)+ NO_3^-(aq)+ K^+(aq)+ HSO_3^-(aq)\rightarrow SO_2(g)+ H_2O(l)+ NO_3^-(aq)+ K^+(aq)$
    Net Ionic: $H^+(aq) + HSO_3^-(aq) \rightarrow SO_2(g) + H_2O(l)$
    d) Complete Ionic:
    $Mn^{3+}(aq)+ 3Cl^-(aq)+ 3K^+(aq)+ PO_4^{3-}(aq)\rightarrow MnPO_4(s)+ 3Cl^-(aq)+ 3K^+(aq)$
    Net Ionic: $Mn^{3+}(aq) + PO_4^{3-}(aq) \rightarrow MnPO_4(s)$

a) acid-base; $KOH(aq) + HC_2H_3O_2(aq) \rightarrow H_2O(l) + KC_2H_3O_2(aq)$
b) gas-evolution/acid-base: $2HBr(aq)+ K_2CO_3(aq)\rightarrow H_2O(l)+ CO_2(g)+ 2KBr(aq)$
) synthesis: $2H_2(g) + O_2(g) \rightarrow 2H_2O(g)$
precipitation: $2NH_4Cl(aq) + Pb(NO_3)_2(aq) \rightarrow PbCl_2(s) + 2NH_4NO_3(aq)$

idation-reduction, single-displacement
-base, gas evolution
volution, double-displacement
tation, double-displacement

101. For calcium chloride:

Molecular: $3CaCl_2(aq) + 2Na_3PO_4(aq) \rightarrow Ca_3(PO_4)_2(s) + 6NaCl(aq)$

Complete: $3Ca^{2+}(aq) + 6Cl^-(aq) + 6Na^+(aq) + 2PO_4^{3-}(aq) \rightarrow Ca_3(PO_4)_2(s) + 6Na^+(aq) + 6Cl^-(aq)$

Net Ionic: $3Ca^{2+}(aq) + 2PO_4^{3-}(aq) \rightarrow Ca_3(PO_4)_2(s)$

For magnesium nitrate:

Molecular: $3Mg(NO_3)_2(aq) + 2Na_3PO_4(aq) \rightarrow Mg_3(PO_4)_2(s) + 6NaNO_3(aq)$

Complete: $3Mg^{2+}(aq) + 6NO_3^-(aq) + 6Na^+(aq) + 2PO_4^{3-}(aq) \rightarrow Mg_3(PO_4)_2(s) + 6Na^+(aq) + 6NO_3^-(aq)$

Net Ionic: $3Mg^{2+}(aq) + 2PO_4^{3-}(aq) \rightarrow Mg_3(PO_4)_2(s)$

103. a) $Pb^{2+}(aq) + 2Cl^-(aq) \rightarrow PbCl_2(s)$

b) $Ca^{2+}(aq) + SO_4^{2-}(aq) \rightarrow CaSO_4(s)$

c) $Ag^+(aq) + Cl^-(aq) \rightarrow AgCl(s)$

d) $Hg_2^{2+}(aq) + 2Cl^-(aq) \rightarrow Hg_2Cl_2(s)$

105. 1) The silver ion is not present as it would precipitate with chloride ions.

2) Sulfate will only precipitate the calcium ion, not the copper(II) ion:

$Ca^{2+}(aq) + SO_4^{2-}(aq) \rightarrow CaSO_4(s)$

3) The only possible ion left is copper(II), which will precipitate with carbonate:

$Cu^{2+}(aq) + CO_3^{2-}(aq) \rightarrow CuCO_3(s)$

The calcium and copper(II) ions were present in the original solution.

107. $2K_3PO_4(aq) + 3Ca^{2+}(aq) \rightarrow Ca_3(PO_4)_2(s) + 6K^+(aq)$

$$0.112 \text{ mol } K_3PO_4 \times \frac{3 \text{ mol } Ca^{2+}}{2 \text{ mol } K_3PO_4} = 0.168 \text{ mol } Ca^{2+}$$

$$0.168 \text{ mol } Ca^{2+} \times \frac{40.08 \text{ g}}{1 \text{ mol } Ca^{2+}} = 6.73 \text{ g } Ca^{2+}$$

109. Assuming lead(II) ions in solution: $Pb^{2+}(aq) + 2NaCl(aq) \rightarrow PbCl_2(s) + 2Na^+(aq)$

$$0.133 \text{ g } Pb^{2+} \times \frac{1 \text{ mol } Pb^{2+}}{207.2 \text{ g}} \times \frac{2 \text{ mol NaCl}}{1 \text{ mol } Pb^{2+}} = 0.00128 \text{ mol NaCl}$$

$$0.133 \text{ g } Pb^{2+} \times \frac{1 \text{mol } Pb^{2+}}{207.2 \text{ g}} \times \frac{2 \text{mol NaCl}}{1 \text{ mol } Pb^{2+}} \times \frac{58.44 \text{ g}}{1 \text{mol NaCl}} = 0.0750 \text{ g NaCl}$$

Highlight Problem

111. The first figure is based on a chemical reaction. You can see the changes in the arrangement of atoms within the compounds before and after detonation.

Quantities in Chemical Reactions 8

Questions

1. The reaction stoichiometry is important because it allows us to calculate how much reactant should be used to form a desired amount of a product. It can be thought of as a recipe for a desired compound. Just think, if we had no way to determine the amount of raw ingredients needed to make a drug in the pharmaceutical industry, there would be no way to produce sufficient amounts at a reasonable cost.

3. $\frac{2 \text{ moles NaCl}}{1 \text{ mole } Cl_2}$

5. mass reactant $\rightarrow$ moles reactant $\rightarrow$ moles product $\rightarrow$ mass product

7. The limiting reactant is the reactant that runs out first and, therefore, determines the maximum amount of product that can be formed in a reaction.

9. The amount of product that is actually produced by a reaction is often less than the theoretical maximum amount. The percent yield of a reaction is the actual yield/theoretical yield $\times$ 100%.

11. d) A and B are present in a mass ratio of 1:2. Since the reaction requires 2 moles of B for every mole of A, then A will be the limiting reagent only if the molar mass of A > molar mass of B. This will result in a molar ratio that is less than 1:2, even though the reactants are present in a 1:2 mass ratio.

13. The enthalpy of reaction (ΔH_{rxn}) is the amount of thermal energy (or heat) that flows when a reaction occurs at constant pressure. This quantity is important as it allows one to calculate the amount of thermal energy produced or consumed by a chemical reaction given a set of specific conditions.

Problems

Mole-to-Mole Conversions

15. a) $2 \text{ mol A} \times \frac{1 \text{ mol C}}{1 \text{ mol A}} = 2 \text{ mol C}$

b) $2 \text{ mol B} \times \frac{1 \text{ mol C}}{2 \text{ mol B}} = 1 \text{ mol C}$

c) $3 \text{ mol A} \times \frac{1 \text{ mol C}}{1 \text{ mol A}} = 3 \text{ mol C}$

d) $3 \text{ mol B} \times \frac{1 \text{ mol C}}{2 \text{ mol B}} = 1.5 \text{ mol C}$

17. a) $1.3 \text{ mol } N_2O_5 \times \frac{4 \text{ mol } NO_2}{2 \text{ mol } N_2O_5} = 2.6 \text{ mol } NO_2$

b) $5.8 \text{ mol } N_2O_5 \times \frac{4 \text{ mol } NO_2}{2 \text{ mol } N_2O_5} = 11.6 \text{ mol } NO_2$

c) $4.45\text{x}10^3 \text{ mol } N_2O_5 \times \frac{4 \text{ mol } NO_2}{2 \text{ mol } N_2O_5} = 8.90x10^3 \text{ mol } NO_2$

d) $1.006\text{x}10^{-3} \text{ mol } N_2O_5 \times \frac{4 \text{ mol } NO_2}{2 \text{ mol } N_2O_5} = 2.012x10^{-3} \text{ mol } NO_2$

19. $2 \text{ molecules } SO_2 \times \frac{2 \text{ molecules } H_2S}{1 \text{ molecule } SO_2} = 4 \text{ molecules } H_2S$ (choice c)

21. a) $1.75 \text{ mol } H_2 \times \frac{2 \text{ mol HCl}}{1 \text{ mol } H_2} = 3.50 \text{ mol HCl}$

b) $1.75 \text{ mol } O_2 \times \frac{2 \text{ mol } H_2O}{1 \text{ mol } O_2} = 3.50 \text{ mol } H_2O$

c) $1.75 \text{ mol Na} \times \frac{1 \text{ mol } Na_2O_2}{2 \text{ mol Na}} = 0.875 \text{ mol } Na_2O_2$

d) $1.75 \text{ mol } O_2 \times \frac{2 \text{ mol } SO_3}{3 \text{ mol } O_2} = 1.17 \text{ mol } SO_3$

23. a) $2.4 \text{ mol PbS} \times \frac{2 \text{ mol PbO}}{2 \text{ mol PbS}} = 2.4 \text{ mol PbO}$

$2.4 \text{ mol PbS} \times \frac{2 \text{ mol } SO_2}{2 \text{ mol PbS}} = 2.4 \text{ mol } SO_2$

b) $2.4 \text{ mol } O_2 \times \frac{2 \text{ mol PbO}}{3 \text{ mol } O_2} = 1.6 \text{ mol PbO}$

$2.4 \text{ mol } O_2 \times \frac{2 \text{ mol } SO_2}{3 \text{ mol } O_2} = 1.6 \text{ mol } SO_2$

c) $5.3 \text{ mol PbS} \times \frac{2 \text{ mol PbO}}{2 \text{ mol PbS}} = 5.3 \text{ mol PbO}$

$5.3 \text{ mol PbS} \times \frac{2 \text{ mol } SO_2}{2 \text{ mol PbS}} = 5.3 \text{ mol } SO_2$

d) $5.3 \text{ mol } O_2 \times \frac{2 \text{ mol PbO}}{3 \text{ mol } O_2} = 3.5 \text{ mol PbO}$

$5.3 \text{ mol } O_2 \times \frac{2 \text{ mol } SO_2}{3 \text{ mol } O_2} = 3.5 \text{ mol } SO_2$

25.

mol N_2H_4	mol N_2O_4	mol N_2	mol H_2O
4	2	6	8
6	3	9	12
4	2	6	8
11	5.5	16.5	22
3	1.5	4.5	6
8.26	4.13	12.4	16.5

27. $2C_4H_{10}(g) + 13O_2(g) \rightarrow 8CO_2(g) + 10\, H_2O(g)$

$4.9 \text{ mol } C_4H_{10} \times \frac{13 \text{ mol } O_2}{2 \text{ mol } C_4H_{10}} = 32 \text{ mol } O_2$

29. a) $Pb(s) + 2AgNO_3(aq) \rightarrow Pb(NO_3)_2(aq) + 2Ag(s)$

b) $9.3 \text{ mol Pb} \times \frac{2 \text{ mol } AgNO_3}{1 \text{ mol Pb}} = 19 \text{ mol } AgNO_3$

c) $28.4 \text{ mol Pb} \times \frac{2 \text{ mol Ag}}{1 \text{ mol Pb}} = 56.8 \text{ mol Ag}$

Mass-to-Mass Conversions

31. a) $2.13 \text{ g HgO} \times \frac{1 \text{ mol HgO}}{216.59 \text{ g}} \times \frac{1 \text{ mol } O_2}{2 \text{ mol HgO}} \times \frac{32.00 \text{ g}}{1 \text{ mol } O_2} = 0.157 \text{ g } O_2$

b) $6.77 \text{ g HgO} \times \frac{1 \text{ mol HgO}}{216.59 \text{ g}} \times \frac{1 \text{ mol } O_2}{2 \text{ mol HgO}} \times \frac{32.00 \text{ g}}{1 \text{ mol } O_2} = 0.500 \text{ g } O_2$

c) $1.55x10^3 \text{ g HgO} \times \frac{1 \text{ mol HgO}}{216.59 \text{ g}} \times \frac{1 \text{ mol } O_2}{2 \text{ mol HgO}} \times \frac{32.00 \text{ g}}{1 \text{ mol } O_2} = 115 \text{ g } O_2$

d) $3.87x10^{-3} \text{ g HgO} \times \frac{1 \text{ mol HgO}}{216.59 \text{ g}} \times \frac{1 \text{ mol } O_2}{2 \text{ mol HgO}} \times \frac{32.00 \text{ g}}{1 \text{ mol } O_2} = 2.86x10^{-4} \text{ g } O_2$

33. a) $2.4 \text{ g } Cl_2 \times \frac{1 \text{ mol } Cl_2}{70.90 \text{ g}} \times \frac{2 \text{ mol NaCl}}{1 \text{ mol } Cl_2} \times \frac{58.44 \text{ g}}{1 \text{ mol NaCl}} = 4.0 \text{ g NaCl}$

b) $2.4 \text{ g CaO} \times \frac{1 \text{ mol CaO}}{56.08 \text{ g}} \times \frac{1 \text{ mol } CaCO_3}{1 \text{ mol CaO}} \times \frac{100.09 \text{ g}}{1 \text{ mol } CaCO_3} = 4.3 \text{ g } CaCO_3$

c) $2.4 \text{ g Mg} \times \frac{1 \text{ mol Mg}}{24.31 \text{ g}} \times \frac{2 \text{ mol MgO}}{2 \text{ mol Mg}} \times \frac{40.31 \text{ g}}{1 \text{ mol MgO}} = 4.0 \text{ g MgO}$

d) $2.4 \text{ g } Na_2O \times \frac{1 \text{ mol } Na_2O}{61.98 \text{ g}} \times \frac{2 \text{ mol NaOH}}{1 \text{ mol } Na_2O} \times \frac{40.00 \text{ g}}{1 \text{ mol NaOH}} = 3.1 \text{ g NaOH}$

35. a) $4.7 \text{ g Al} \times \frac{1 \text{ mol Al}}{26.98 \text{ g}} \times \frac{1 \text{ mol } Al_2O_3}{2 \text{ mol Al}} \times \frac{101.96 \text{ g}}{1 \text{ mol } Al_2O_3} = 8.9 \text{ g } Al_2O_3$

$4.7 \text{ g Al} \times \frac{1 \text{ mol Al}}{26.98 \text{ g}} \times \frac{2 \text{ mol Fe}}{2 \text{ mol Al}} \times \frac{55.85 \text{ g}}{1 \text{ mol Fe}} = 9.7 \text{ g Fe}$

b) $4.7 \text{ g } Fe_2O_3 \times \frac{1 \text{ mol } Fe_2O_3}{159.70 \text{ g}} \times \frac{1 \text{ mol } Al_2O_3}{1 \text{ mol } Fe_2O_3} \times \frac{101.96 \text{ g}}{1 \text{ mol } Al_2O_3} = 3.0 \text{ g } Al_2O_3$

$4.7 \text{ g } Fe_2O_3 \times \frac{1 \text{ mol } Fe_2O_3}{159.70 \text{ g}} \times \frac{2 \text{ mol Fe}}{1 \text{ mol } Fe_2O_3} \times \frac{55.85 \text{ g}}{1 \text{ mol Fe}} = 3.3 \text{ g Fe}$

37.

Mass CH_4	Mass O_2	Mass CO_2	Mass H_2O
0.645 g	2.57 g	1.77 g	1.45 g
22.32 g	89.00 g	61.20 g	50.12 g
5.041 g	20.10 g	13.82 g	11.32 g
1.07 g	4.28 g	2.94 g	2.41 g
3.18 kg	12.7 kg	8.72 kg	7.14 kg
$8.57x10^2$ kg	$3.42x10^3$ kg	$2.35x10^3$ kg	$1.92x10^3$ kg

39. a) $2.5 \text{ g NaOH} \times \frac{1 \text{ mol NaOH}}{40.00 \text{ g}} \times \frac{1 \text{ mol HCl}}{1 \text{ mol NaOH}} \times \frac{36.46 \text{ g}}{1 \text{ mol HCl}} = 2.3 \text{ g HCl}$

b) $2.5 \text{ g Ca(OH)}_2 \times \frac{1 \text{ mol Ca(OH)}_2}{74.10 \text{ g}} \times \frac{2 \text{ mol HNO}_3}{1 \text{ mol Ca(OH)}_2} \times \frac{63.02 \text{ g}}{1 \text{ mol HNO}_3}$

$= 4.3 \text{ g HNO}_3$

c) $2.5 \text{ g KOH} \times \frac{1 \text{ mol KOH}}{56.11 \text{ g}} \times \frac{1 \text{ mol H}_2\text{SO}_4}{2 \text{ mol KOH}} \times \frac{98.09 \text{ g}}{1 \text{ mol H}_2\text{SO}_4} = 2.2 \text{ g H}_2\text{SO}_4$

41. $22.5 \text{ g Al} \times \frac{1 \text{ mol Al}}{26.98 \text{ g}} \times \frac{3 \text{ mol H}_2\text{SO}_4}{2 \text{ mol Al}} \times \frac{98.09 \text{ g}}{1 \text{ mol H}_2\text{SO}_4} = 123 \text{ g H}_2\text{SO}_4$

$22.5 \text{ g Al} \times \frac{1 \text{ mol Al}}{26.98 \text{ g}} \times \frac{3 \text{ mol H}_2}{2 \text{ mol Al}} \times \frac{2.02 \text{ g}}{1 \text{ mol H}_2} = 2.53 \text{ g H}_2$

Limiting Reactant, Theoretical Yield, and Percent Yield

43. a) $2 \text{ mol A} \times \frac{3 \text{ mol C}}{2 \text{ mol A}} = 3 \text{ mol C}$

$5 \text{ mol B} \times \frac{3 \text{ mol C}}{4 \text{ mol B}} = 3.75 \text{ mol C}$ $\therefore$ The limiting reactant is A.

b) $1.8 \text{ mol A} \times \frac{3 \text{ mol C}}{2 \text{ mol A}} = 2.7 \text{ mol C}$

$4 \text{ mol B} \times \frac{3 \text{ mol C}}{4 \text{ mol B}} = 3 \text{ mol C}$ $\therefore$ The limiting reactant is A.

c) $3 \text{ mol A} \times \frac{3 \text{ mol C}}{2 \text{ mol A}} = 4.5 \text{ mol C}$

$4 \text{ mol B} \times \frac{3 \text{ mol C}}{4 \text{ mol B}} = 3 \text{ mol C}$ $\therefore$ The limiting reactant is B.

d) $22 \text{ mol A} \times \frac{3 \text{ mol C}}{2 \text{ mol A}} = 33 \text{ mol C}$

$40 \text{ mol B} \times \frac{3 \text{ mol C}}{4 \text{ mol B}} = 30 \text{ mol C}$ $\therefore$ The limiting reactant is B.

45. a) $1 \text{ mol A} \times \frac{3 \text{ mol C}}{1 \text{ mol A}} = 3 \text{ mol C}$

$1 \text{ mol B} \times \frac{3 \text{ mol C}}{2 \text{ mol B}} = 1.5 \text{ mol C}$

The limiting reactant is B. Therefore, the theoretical yield of C is 1.5 mol

b) $2 \text{ mol A} \times \dfrac{3 \text{ mol C}}{1 \text{ mol A}} = 6 \text{ mol C}$

$2 \text{ mol B} \times \dfrac{3 \text{ mol C}}{2 \text{ mol B}} = 3 \text{ mol C}$

The limiting reactant is B. Therefore, the theoretical yield of C is 3 mol.

c) $1 \text{ mol A} \times \dfrac{3 \text{ mol C}}{1 \text{ mol A}} = 3 \text{ mol C}$

$3 \text{ mol B} \times \dfrac{3 \text{ mol C}}{2 \text{ mol B}} = 4.5 \text{ mol C}$

The limiting reactant is A. Therefore, the theoretical yield of C is 3 mol.

d) $32 \text{ mol A} \times \dfrac{3 \text{ mol C}}{1 \text{ mol A}} = 96 \text{ mol C}$

$68 \text{ mol B} \times \dfrac{3 \text{ mol C}}{2 \text{ mol B}} = 102 \text{ mol C}$

The limiting reactant is A. Therefore, the theoretical yield of C is 96 mol.

47. a) $1 \text{ mol K} \times \dfrac{2 \text{ mol KCl}}{2 \text{ mol K}} = 1 \text{ mol KCl}$

$1 \text{ mol Cl}_2 \times \dfrac{2 \text{ mol KCl}}{1 \text{ mol Cl}_2} = 2 \text{ mol KCl}$ $\therefore$ The limiting reactant is K.

b) $1.8 \text{ mol K} \times \dfrac{2 \text{ mol KCl}}{2 \text{ mol K}} = 1.8 \text{ mol KCl}$

$1 \text{ mol Cl}_2 \times \dfrac{2 \text{ mol KCl}}{1 \text{ mol Cl}_2} = 2 \text{ mol KCl}$ $\therefore$ The limiting reactant is K.

c) $2.2 \text{ mol K} \times \dfrac{2 \text{ mol KCl}}{2 \text{ mol K}} = 2.2 \text{ mol KCl}$

$1 \text{ mol Cl}_2 \times \dfrac{2 \text{ mol KCl}}{1 \text{ mol Cl}_2} = 2 \text{ mol KCl}$ $\therefore$ The limiting reactant is Cl_2.

d) $14.6 \text{ mol K} \times \dfrac{2 \text{ mol KCl}}{2 \text{ mol K}} = 14.6 \text{ mol KCl}$

$7.8 \text{ mol Cl}_2 \times \dfrac{2 \text{ mol KCl}}{1 \text{ mol Cl}_2} = 15.6 \text{ mol KCl}$ $\therefore$ The limiting reactant is K.

49. a) $2 \text{ mol Mn} \times \frac{2 \text{ mol } MnO_3}{2 \text{ mol Mn}} = 2 \text{ mol } MnO_3$

$2 \text{ mol } O_2 \times \frac{2 \text{ mol } MnO_3}{3 \text{ mol } O_2} = 1.3 \text{ mol } MnO_3$

The limiting reactant is O_2, $\therefore$ the theoretical yield of MnO_3 is 1.3 mol.

b) $4.8 \text{ mol Mn} \times \frac{2 \text{ mol } MnO_3}{2 \text{ mol Mn}} = 4.8 \text{ mol } MnO_3$

$8.5 \text{ mol } O_2 \times \frac{2 \text{ mol } MnO_3}{3 \text{ mol } O_2} = 5.7 \text{ mol } MnO_3$

The limiting reactant is Mn, $\therefore$ the theoretical yield of MnO_3 is 4.8 mol.

c) $0.114 \text{ mol Mn} \times \frac{2 \text{ mol } MnO_3}{2 \text{ mol Mn}} = 0.114 \text{ mol } MnO_3$

$0.161 \text{ mol } O_2 \times \frac{2 \text{ mol } MnO_3}{3 \text{ mol } O_2} = 0.107 \text{ mol } MnO_3$

The limiting reactant is O_2, $\therefore$ the theoretical yield of MnO_3 is 0.107 mol.

d) $27.5 \text{ mol Mn} \times \frac{2 \text{ mol } MnO_3}{2 \text{ mol Mn}} = 27.5 \text{ mol } MnO_3$

$43.8 \text{ mol } O_2 \times \frac{2 \text{ mol } MnO_3}{3 \text{ mol } O_2} = 29.2 \text{ mol } MnO_3$

The limiting reactant is Mn, $\therefore$ the theoretical yield of MnO_3 is 27.5 mol.

51. $9 \text{ mol A} \times \frac{2 \text{ mol C}}{3 \text{ mol A}} = 6 \text{ mol C}$

$8 \text{ mol B} \times \frac{2 \text{ mol C}}{4 \text{ mol B}} = 4 \text{ mol C}$ $\therefore$ B is the limiting reagent

Final reaction vessel would contain:

4 mol C (theoretical yield)

0 mol B (the limiting reagent)

Excess mol A = Starting moles - Consumed moles

Starting moles = 9 moles A

Consumed moles $= 8 \text{ mol B} \times \frac{3 \text{ mol A}}{4 \text{ mol B}} = 6 \text{ mol A}$

Excess mol A = 9 mol A - 6 mol A= 3 mol A

53. a) Because there is only one O_2 molecule and it only requires four of the seven available HCl molecules, two Cl_2 molecules are formed.
b) There are only six available HCl molecules, however, twelve molecules are needed to fully react with the three available O_2 molecules. Therefore, only three Cl_2 molecules are formed.
c) There are only four HCl molecules available and excess O_2 molecules, therefore two Cl_2 molecules are formed.

55. a) $1.0\text{ g Li} \times \frac{1\text{ mol Li}}{6.94\text{ g}} \times \frac{2\text{ mol LiF}}{2\text{ mol Li}} = 0.14\text{ mol LiF}$

$1.0\text{ g }F_2 \times \frac{1\text{ mol }F_2}{38.00\text{ g}} \times \frac{2\text{ mol LiF}}{1\text{ mol }F_2} = 0.053\text{ mol LiF}$ ∴ Limiting reactant is F_2.

b) $10.5\text{ g Li} \times \frac{1\text{ mol Li}}{6.94\text{ g}} \times \frac{2\text{ mol LiF}}{2\text{ mol Li}} = 1.51\text{ mol LiF}$

$37.2\text{ g }F_2 \times \frac{1\text{ mol }F_2}{38.00\text{ g}} \times \frac{2\text{ mol LiF}}{1\text{ mol }F_2} = 1.96\text{ mol LiF}$ ∴ Limiting reactant is Li.

c) $2.85x10^3\text{g Li} \times \frac{1\text{ mol Li}}{6.94\text{ g}} \times \frac{2\text{ mol LiF}}{2\text{ mol Li}} = 411\text{ mol LiF}$

$6.79x10^3\text{g }F_2 \times \frac{1\text{ mol }F_2}{38.00\text{ g}} \times \frac{2\text{ mol LiF}}{1\text{ mol }F_2} = 357\text{ mol LiF}$ ∴ Limiting reactant is F_2.

57. a) $1.0\text{ g Al} \times \frac{1\text{ mol Al}}{26.98\text{ g}} \times \frac{2\text{ mol }AlCl_3}{2\text{ mol Al}} \times \frac{133.33\text{ g}}{1\text{ mol }AlCl_3} = 4.9\text{ g }AlCl_3$

$1.0\text{ g }Cl_2 \times \frac{1\text{ mol }Cl_2}{70.90\text{ g}} \times \frac{2\text{ mol }AlCl_3}{3\text{ mol }Cl_2} \times \frac{133.33\text{ g}}{1\text{ mol }AlCl_3} = 1.3\text{ g }AlCl_3$

The limiting reactant is Cl_2, the theoretical yield is 1.3 g of $AlCl_3$.

b) $5.5\text{ g Al} \times \frac{1\text{ mol Al}}{26.98\text{ g}} \times \frac{2\text{ mol }AlCl_3}{2\text{ mol Al}} \times \frac{133.33\text{ g}}{1\text{ mol }AlCl_3} = 27\text{ g }AlCl_3$

$19.8\text{ g }Cl_2 \times \frac{1\text{ mol }Cl_2}{70.90\text{ g}} \times \frac{2\text{ mol }AlCl_3}{3\text{ mol }Cl_2} \times \frac{133.33\text{ g}}{1\text{ mol }AlCl_3} = 24.8\text{ g }AlCl_3$

The limiting reactant is Cl_2, the theoretical yield is 24.8 g of $AlCl_3$.

c) $0.439\text{ g Al} \times \frac{1\text{ mol Al}}{26.98\text{ g}} \times \frac{2\text{ mol }AlCl_3}{2\text{ mol Al}} \times \frac{133.33\text{ g}}{1\text{ mol }AlCl_3} = 2.17\text{ g }AlCl_3$

$2.29\text{ g }Cl_2 \times \frac{1\text{ mol }Cl_2}{70.90\text{ g}} \times \frac{2\text{ mol }AlCl_3}{3\text{ mol }Cl_2} \times \frac{133.33\text{ g}}{1\text{ mol }AlCl_3} = 2.87\text{ g }AlCl_3$

The limiting reactant is Al, the theoretical yield is 2.17 g $AlCl_3$.

59. Percent Yield$=\frac{18.5}{24.8}\times 100\% = 74.6\%$

61. $14.4\text{ g CaO}\times\frac{1\text{ mol CaO}}{56.08\text{ g}}\times\frac{1\text{ mol CaCO}_3}{1\text{ mol CaO}}\times\frac{100.09\text{ g}}{1\text{ mol CaCO}_3}=25.7\text{ g CaCO}_3$

$13.8\text{ g CO}_2\times\frac{1\text{ mol CO}_2}{44.01\text{ g}}\times\frac{1\text{ mol CaCO}_3}{1\text{ mol CO}_2}\times\frac{100.09\text{ g}}{1\text{ mol CaCO}_3}=31.4\text{ g CaCO}_3$

The limiting reactant is CaO.

The theoretical yield is 25.7 g $CaCO_3$.

The percent yield$=\frac{19.4}{25.7}\times 100\% = 75.5\%$

63. $11.2\text{ g NiS}_2\times\frac{1\text{ mol NiS}_2}{122.83\text{ g}}\times\frac{2\text{ mol NiO}}{2\text{ mol NiS}_2}\times\frac{74.69\text{ g}}{1\text{ mol NiO}}=6.81\text{ g NiO}$

$5.43\text{ g O}_2\times\frac{1\text{ mol O}_2}{32.00\text{ g}}\times\frac{2\text{ mol NiO}}{5\text{ mol O}_2}\times\frac{74.69\text{ g}}{1\text{ mol NiO}}=5.07\text{ g NiO}$

The limiting reactant is O_2.

The theoretical yield is 5.07 g NiO.

The percent yield$=\frac{4.86}{5.07}\times 100\% = 95.9\%$

65. $135.8\text{ g NaCl}\times\frac{1\text{ mol NaCl}}{58.44\text{ g}}\times\frac{1\text{ mol PbCl}_2}{2\text{ mol NaCl}}\times\frac{278.1\text{ g}}{1\text{ mol PbCl}_2}=323.1\text{ g PbCl}_2$

$195.7\text{ g Pb}^{2+}\times\frac{1\text{ mol Pb}^{2+}}{207.2\text{ g}}\times\frac{1\text{ mol PbCl}_2}{1\text{ mol Pb}^{2+}}\times\frac{278.1\text{ g}}{1\text{ mol PbCl}_2}=262.7\text{ g PbCl}_2$

The limiting reactant is Pb^{2+}.

The theoretical yield is 262.7 g $PbCl_2$.

The percent yield$=\frac{252.4}{262.7}\times 100\% = 96.1\%$

67. $10.0\text{ g TiO}_2 \times \frac{1\text{ mol TiO}_2}{79.87\text{ g}} \times \frac{1\text{ mol Ti}}{1\text{ mole TiO}_2} \times \frac{47.87\text{ g}}{1\text{ mol Ti}} = 5.99\text{ g Ti}$

$10.0\text{ g C} \times \frac{1\text{ mol C}}{12.01\text{ g}} \times \frac{1\text{ mol Ti}}{2\text{ mole C}} \times \frac{47.87\text{ g}}{1\text{ mol Ti}} = 19.9\text{ g Ti}$

$\therefore$ C is excess reagent, $TiO_2$is the limiting reagent

Upon completion the reaction vessel would contain

0 g TiO_2 (the limiting reagent)

5.99 g Ti

$10.0\text{ g TiO}_2 \times \frac{1\text{ mol TiO}_2}{79.866\text{ g}} \times \frac{2\text{ mol CO}}{1\text{ mole TiO}_2} \times \frac{28.01\text{ g}}{1\text{ mol CO}} = 7.01\text{ g CO}$

Mass of unused carbon can be calculated using the Law of conservation of mass

20.0 g of reactants, 13.0 g products (5.99 g Ti + 7.01 g CO), therefore 7.00 g carbon left

Enthalpy and Stoichiometry ΔH_{rxn}

69. a) exothermic, $-\Delta H$
b) endothermic, $+\Delta H$
c) exothermic, $-\Delta H$

71. a) $1\text{ mol A} \times \frac{-55\text{ kJ}}{1\text{ mol A}} = -55\text{ kJ}$

b) $2\text{ mol A} \times \frac{-55\text{ kJ}}{1\text{ mol A}} = -110\text{ kJ} \Rightarrow -1.1\text{x}10^2\text{kJ}$

c) $1\text{ mol B} \times \frac{-55\text{ kJ}}{2\text{ mol B}} = -27.5\text{ kJ} \Rightarrow -28\text{kJ}$

d) $2\text{ mol B} \times \frac{-55\text{ kJ}}{2\text{ mol B}} = -55\text{ kJ}$

73. $155\text{ g C}_3\text{H}_6\text{O} \times \frac{1\text{ mol C}_3\text{H}_6\text{O}}{58.09\text{ g C}_3\text{H}_6\text{O}} \times \frac{-1790\text{ kJ}}{1\text{ mol C}_3\text{H}_6\text{O}} = -4.78 \times 10^3\text{ kJ}$

75. $-1.55\text{x}10^3\text{ kJ} \times \frac{1\text{ mol C}_8\text{H}_{18}}{-5074.1\text{ kJ}} \times \frac{114.26\text{ g}}{1\text{ mol C}_8\text{H}_{18}} = 34.9\text{ g C}_8\text{H}_{18}$

Cumulative Problems

77. You can estimate about how much of each reactant is present: approximately 1 mole N_2, 4-5 mole O_2 and 2 mole of H_2O. Looking at the reaction you need 2 mole of N_2 to 5 mole of O_2 to 2 mole of H_2O, therefore N_2 is the limiting reagent.

79. $Ba^{2+}(aq) + Na_2SO_4(aq) \rightarrow BaSO_4(s) + 2Na^{+}(aq)$

$$258\text{ mg BaSO}_4 \times \frac{1\text{ g}}{1000\text{ mg}} \times \frac{1\text{ mol BaSO}_4}{233.40\text{ g}} \times \frac{1\text{ mol Ba}^{2+}}{1\text{ mol BaSO}_4} \times \frac{137.33\text{ g}}{1\text{ mol Ba}^{2+}}$$

$= 0.152\text{ g Ba}^{2+}$

81. $NaHCO_3(aq) + HCl(aq) \rightarrow H_2O(l) + CO_2(g) + NaCl(aq)$

$$3.5\text{ g NaHCO}_3 \times \frac{1\text{ mol NaHCO}_3}{84.01\text{ g}} \times \frac{1\text{ mol HCl}}{1\text{ mol NaHCO}_3} \times \frac{36.46\text{ g}}{1\text{ mol HCl}} = 1.5\text{ g HCl}$$

83. $2C_8H_{18}(l) + 25O_2(g) \rightarrow 18H_2O(l) + 16CO_2(g)$

$$1.0\text{ kg C}_8\text{H}_{18} \times \frac{1\times10^3\text{g}}{1\text{ kg}} \times \frac{1\text{ mol C}_8\text{H}_{18}}{114.26\text{ g}} \times \frac{16\text{ mol CO}_2}{2\text{ mol C}_8\text{H}_{18}} \times \frac{44.01\text{ g}}{1\text{ mol CO}_2} \times \frac{1\text{ kg}}{1\times10^3\text{g}}$$

$= 3.1\text{ kg CO}_2$

85. $3CaCl_2(aq) + 2Na_3PO_4(aq) \rightarrow Ca_3(PO_4)_2(s) + 6NaCl(aq)$

$$4.8\text{ g CaCl}_2 \times \frac{1\text{ mol CaCl}_2}{110.98\text{ g}} \times \frac{2\text{ mol Na}_3\text{PO}_4}{3\text{ mol CaCl}_2} \times \frac{163.94\text{ g}}{1\text{ mol Na}_3\text{PO}_4} = 4.7\text{ g Na}_3\text{PO}_4$$

87. $Zn(s) + 2HCl(aq) \rightarrow H_2(g) + ZnCl_2\ (aq)$

$$14.5\text{ g H}_2 \times \frac{1\text{ mol H}_2}{2.02\text{ g}} \times \frac{1\text{ mol Zn}}{1\text{ mol H}_2} \times \frac{65.39\text{ g}}{1\text{ mol Zn}} = 4.69x10^2\text{ g Zn}$$

89. $2NH_4NO_3(s) \rightarrow 2N_2(g) + O_2(g) + 4H_2O(g)$

$$1.00\times10^3\text{gNH}_4\text{NO}_3 \times \frac{1\text{ mol NH}_4\text{NO}_3}{80.06\text{ g}} \times \frac{1\text{ mol O}_2}{2\text{ mol NH}_4\text{NO}_3} \times \frac{32.00\text{ g}}{1\text{ mol O}_2} = 2.00\times10^2\text{g O}_2$$

91. $$5.00\text{ mL C}_4\text{H}_6\text{O}_3 \times \frac{1.08\text{ g}}{1\text{ mL}} \times \frac{1\text{ mol C}_4\text{H}_6\text{O}_3}{102.10\text{ g}} \times \frac{1\text{ mol C}_9\text{H}_8\text{O}_4}{1\text{ mol C}_4\text{H}_6\text{O}_3} \times \frac{180.17\text{ g}}{1\text{ mol C}_9\text{H}_8\text{O}_4}$$

$= 9.53\text{ g C}_9\text{H}_8\text{O}_4$

$$2.08\text{ g C}_7\text{H}_6\text{O}_3 \times \frac{1\text{ mol C}_7\text{H}_6\text{O}_3}{138.13\text{ g}} \times \frac{1\text{ mol C}_9\text{H}_8\text{O}_4}{1\text{ mol C}_7\text{H}_6\text{O}_3} \times \frac{180.17\text{ g}}{1\text{ mol C}_9\text{H}_8\text{O}_4} = 2.71\text{ g C}_9\text{H}_8\text{O}_4$$

The limiting reagent is salicylic acid, $C_7H_6O_3$.

The theoretical yield of aspirin, $C_9H_8O_4$, is then 2.71 g.

The percent yield$= \frac{2.01}{2.71} \times 100\% = 74.2\%$

93. $68.2 \text{ kg NH}_3 \times \frac{1\times10^3\text{g}}{1\text{ kg}} \times \frac{1\text{ mol NH}_3}{17.04\text{ g}} \times \frac{1\text{ mol CH}_4\text{N}_2\text{O}}{2\text{ mol NH}_3} \times \frac{60.07\text{ g}}{1\text{ mol CH}_4\text{N}_2\text{O}} \times \frac{1\text{ kg}}{1000\text{g}}$

$=1.20\times10^2 \text{ kg CH}_4\text{N}_2\text{O}$

$105 \text{ kg CO}_2 \times \frac{1\times10^3\text{g}}{1\text{ kg}} \times \frac{1\text{ mol CO}_2}{44.01\text{ g}} \times \frac{1\text{ mol CH}_4\text{N}_2\text{O}}{1\text{ mol CO}_2} \times \frac{60.07\text{ g}}{1\text{ mol CH}_4\text{N}_2\text{O}} \times \frac{1\text{ kg}}{1000\text{g}}$

$=143 \text{ kg CH}_4\text{N}_2\text{O}$

The limiting reagent is ammonia, NH_3.

The theoretical yield of urea is then 1.20×10^2 kg.

The percent yield$=\frac{87.5}{120}\times100\% = 72.9\%$

95. $\frac{0.550\text{ mg Pb}}{\text{L}} \times \frac{1\text{ g}}{1000\text{ mg}} \times 5.0\text{ L} \times \frac{1\text{ mol Pb}}{207.2\text{ g}} \times \frac{1\text{ mol C}_4\text{H}_6\text{O}_4\text{S}_2}{1\text{ mol Pb}} \times \frac{182.2\text{ g}}{1\text{ mol C}_4\text{H}_6\text{O}_4\text{S}_2}$

$= 0.00242$ g or 2.42 mg of Succimer

97. If cooking is 10% efficient, then 1.6×10^3 kJ $\times 10 = 1.6\times10^4$ kJ heat are released.

$-1.6\times10^4 \text{ kJ} \times \frac{3\text{ mol CO}_2}{-2044\text{ kJ}} \times \frac{44.01\text{ g}}{1\text{ mol CO}_2} = 1.0\times10^3\text{g CO}_2$

Highlight Problems

99. The balloon in (b) has a 2:1 ratio of reactants that matches the reaction stoichiometry.

101. Solution is based on the assumption all numbers have 3 significant digits (this was not stated in the text of the problem).

$2\text{C}_8\text{H}_{18}(l) + 25\text{O}_2(g) \rightarrow 18\text{H}_2\text{O}(l) + 16\text{CO}_2(g)$

$9.0\times10^{12}\text{kg C}_8\text{H}_{18} \times \frac{1\times10^3\text{g}}{1\text{ kg}} \times \frac{1\text{ mol C}_8\text{H}_{18}}{114.26\text{ g}} \times \frac{16\text{ mol CO}_2}{2\text{ mol C}_8\text{H}_{18}} \times \frac{44.01\text{ g}}{1\text{ mol CO}_2} \times \frac{1\text{ kg}}{1000\text{g}}$

$= 2.77\times10^{13} \text{ kg CO}_2/\text{year}$

$(2.77\times10^{13}\text{kg CO}_2/\text{year})(\text{X years}) = 3.00\times10^{15}\text{kg CO}_2 \Rightarrow$

$\text{X}=1.10\times10^2$ years

Electrons in Atoms and the Periodic Table

9

Questions

1. Both the Bohr model and the quantum-mechanical model were developed in the early 1900s. These models serve to explain how electrons are arranged within the atomic structure and how the electrons affect the chemical and physical properties of each element.

3. White light actually contains the wavelengths of many colors (red, orange, yellow, green, blue, indigo, and violet). We see a certain color of light when only the specific wavelength which corresponds to that color is present.

5. The wavelength of light and energy are inversely related, the shorter the wavelength the greater the energy. Wavelength and frequency are inversely related—the shorter the wavelength, the higher the frequency.

7. X-rays are used to image internal bones and organs.

9. The ultraviolet rays, which cause sunburn and suntan, also have the ability to damage biological molecules. UV rays are linked to skin cancer, cataracts, and premature skin wrinkling.

11. Only certain molecules, such as water, can absorb microwaves. This explains why your food, which contains water, is heated while your plate, which doesn't contain water, does not heat.

13. The Bohr model for hydrogen places the single electron in a circular orbit around the nucleus. The orbit of the electron is quantized. That is, it has a fixed energy at a specific fixed distance from the nucleus.

15. The Bohr model orbit is a circular orbit that maps the exact path an electron would make around a nucleus. The quantum mechanical model however has an orbital that is best described as a region that has the greatest probability for the location of an electron.

17. The e^- has wave-particle duality which means the path of an electron is not predictable. The motion of a baseball is predictable. A probability map shows a statistical, reproducible pattern of where the electron is located.

19. The four subshells are *s*, *p*, *d*, and *f*. The maximum number of electrons are 2, 6, 10, and 14, respectively.

21. The Pauli exclusion principle says that the maximum number of electrons in an orbital is two. The two electrons must have opposite spin direction to occupy the same orbital. This is important in assigning electron configurations, as it allows you to determine how and where the electrons should be assigned.

23. [Ne] represents $1s^22s^22p^6$
[Kr] represents $1s^22s^22p^63s^23p^64s^23d^{10}4p^6$

25.

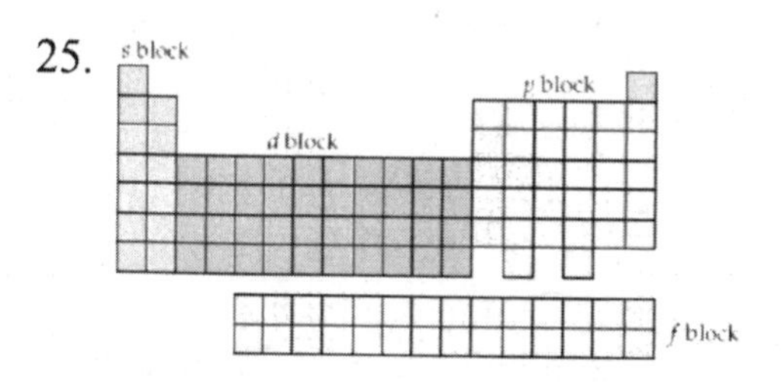

27. Group 1 elements form +1 ions because the electron configuration of the ion will match that of a noble gas. The group 7 elements form -1 ions for the same reason.

Problems

Wavelength, Energy, and Frequency of Electromagnetic Radiation

29. a) $1.0 \text{ ft} \times \frac{12 \text{ in}}{1 \text{ ft}} \times \frac{2.54 \text{ cm}}{1 \text{ in}} \times \frac{1 \text{ m}}{100 \text{cm}} \times \frac{1 \text{ s}}{3.00 \times 10^8 \text{m}} \times \frac{1 \text{ ns}}{1 \times 10^{-9} \text{s}} = 1.0 \text{ ns}$

b) $2462 \text{ mi} \times \frac{5280 \text{ ft}}{1 \text{ mi}} \times \frac{12 \text{ in}}{1 \text{ ft}} \times \frac{2.54 \text{ cm}}{1 \text{ in}} \times \frac{1 \text{ m}}{100 \text{ cm}} \times \frac{1 \text{ s}}{3.00 \times 10^8 \text{m}} \times \frac{1 \text{ ms}}{1 \times 10^{-3} \text{s}} = 13.21 \text{ ms}$

c) $4.5 \times 10^9 \text{km} \times \frac{1000 \text{ m}}{1 \text{ km}} \times \frac{1 \text{ s}}{3.00 \times 10^8 \text{m}} \times \frac{1 \text{ min}}{60 \text{s}} \times \frac{1 \text{ hr}}{60 \text{ min}} = 4.2 \text{ hr (4hr 10min)}$

31. c) infrared

33. radio waves < microwaves < infrared < ultraviolet

35. Two of the following three: gamma rays, X-rays, or ultraviolet rays.

37. a) radio waves < infrared < X-rays
b) radio waves < infrared < X-rays
c) X-rays < infrared < radio waves

The Bohr Model

39. Bohr orbits have fixed <u>energies</u> and fixed <u>distances</u>.

41. The smaller the wavelength the larger the energy of the photon. Therefore, the 410nm wavelength corresponds to the n=6 to n=2 transition and the 434nm wavelength corresponds to the n=5 to n=2 transition.

The Quantum-mechanical Model

43. The 2s and 3p would have the same shape as the 1s and 2p. The only difference is that they would be larger in size.

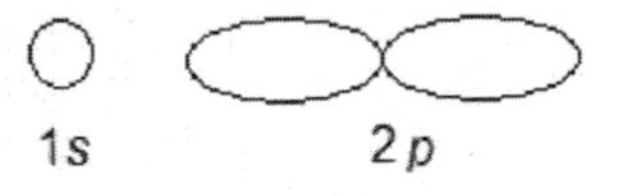

45. On average, the 2s e^- is closer to the nucleus, because it is in a smaller orbital.

47. The transition with the smallest energy difference will produce the longer wavelength. This would correspond to the 2p to 1s transition.

Electron Configurations

49. a) Sr: $1s^2 2s^2 2p^6 3s^2 3p^6 4s^2 3d^{10} 4p^6 5s^2$
b) Ge: $1s^2 2s^2 2p^6 3s^2 3p^6 4s^2 3d^{10} 4p^2$
c) Li: $1s^2 2s^1$
d) Kr: $1s^2 2s^2 2p^6 3s^2 3p^6 4s^2 3d^{10} 4p^6$

51. a) He: [↑↓] 0 unpaired
(1s)

b) B: [↑↓] [↑↓] [↑][ ][ ] 1 unpaired
(1s, 2s, 2p)

c) Li: [↑↓] [↑] 1 unpaired
(1s, 2s)

d) N: [↑↓] [↑↓] [↑][↑][↑] 3 unpaired
(1s, 2s, 2p)

53. a) $[Ar]4s^2 3d^{10} 4p^1$
b) $[Ar]4s^2 3d^{10} 4p^3$
c) $[Kr]5s^1$
d) $[Kr]5s^2 4d^{10} 5p^2$

55. a) Zn: $[Ar]4s^2 3d^{10}$
b) Cu: $[Ar]4s^1 3d^{10}$
c) Zr: $[Kr]5s^2 4d^2$
d) Fe: $[Ar]4s^2 3d^6$

Valence Electrons and Core Electrons

57. Valence electrons are <u>underlined</u>.
    a) Kr: $1s^22s^22p^63s^23p^6\underline{4s^2}3d^{10}\underline{4p^6}$
    b) Ge: $1s^22s^22p^63s^23p^6\underline{4s^2}3d^{10}\underline{4p^2}$
    c) Cl: $1s^22s^22p^6\underline{3s^23p^5}$
    d) Sr: $1s^22s^22p^63s^23p^64s^23d^{10}4p^6\underline{5s^2}$

59. a) Br: [↑↓] (4s) [↑↓][↑↓][↑] (4p) 1 unpaired electron
    b) Kr: [↑↓] (4s) [↑↓][↑↓][↑↓] (4p) 0 unpaired electron
    c) Na: [↑] (3s) [ ][ ][ ] (3p) 1 unpaired electron
    d) In: [↑↓] (5s) [↑][ ][ ] (5p) 1 unpaired electron

61. a) 6
    b) 6
    c) 7
    d) 1

Electron Configurations and the Periodic Table

63. a) ns^1
    b) ns^2
    c) ns^2np^3
    d) ns^2np^5

65. a) $[Ne]3s^23p^1$
    b) $[He]2s^2$
    c) $[Kr]5s^24d^{10}5p^1$
    d) $[Kr]5s^24d^2$

67. a) Sr: $[Kr]5s^2$
    b) Y: $[Kr]5s^24d^1$
    c) Ti: $[Ar]4s^23d^2$
    d) Te: $[Kr]5s^24d^{10}5p^4$

69. a) 2
    b) 3
    c) 5
    d) 6

71. Period 1 consists of two elements within the 1s subshell. The *s* subshell has a maximum of two elements. Period 2 consists of eight elements within the 2*s* and 2*p* subshells. The *s* subshell has a maximum of two elements and the *p* subshell has a maximum of six elements for a combined total of 8 elements.

73. a) aluminum
 b) sulfur
 c) argon
 d) magnesium

75. a) chlorine
 b) gallium
 c) iron
 d) rubidium

Periodic Trends

77. a) As
 b) Br
 c) cannot determine based on periodic properties alone
 d) S

79. Pb < Sn < Te < S < Cl

81. a) In
 b) Si
 c) Pb
 d) C

83. F < S < Si < Ge < Ca < Rb

85. a) Sr
 b) Bi
 c) cannot determine based on periodic properties alone
 d) As

87. S < Se < Sb < In < Ba < Fr

Cumulative Problems

89. When n=3, there can be 3*s*, 3*p*, and 3*d* subshells. The *s* has 2 e^-, *p* has 6 e^- and the *d* has 10 e^- for a total of 18 electrons.

91. The alkali metals all share the [ ]ns^2 electron configuration. By losing two electrons to form the 2+ ion, the electron configuration of the ion becomes the same as a noble gas.

93. a) $1s^2 2s^2 2p^6 3s^2 3p^6$
b) $1s^2 2s^2 2p^6 3s^2 3p^6$
c) $1s^2 2s^2 2p^6 3s^2 3p^6$
d) $1s^2 2s^2 2p^6 3s^2 3p^6 4s^2 3d^{10} 4p^6$
All electron configurations are isoelectronic with noble gases.

95. Metals are on the left side of the periodic table because they tend to give up electrons to form positive ions, which assume an electron configuration identical to that of a noble gas. Nonmetals are on the right side of the periodic table because they tend to gain electrons to form negative ions, which assume an electron configuration identical to that of a noble gas. The metalloids are the boundary region between the metals and nonmetals.

97. a) There is a maximum of 6 *p* and 2 *s* electrons for any principal quantum number.
$1s^2 2s^2 2p^6 3s^2 3p^3$
b) There is not a 2*d* subshell.
$1s^2 2s^2 2p^6 3s^2 3p^2$
c) There is not a 1*p* subshell.
$1s^2 2s^2 2p^3$
d) There is a maximum of 6 *p* electrons for any principle quantum number.
$1s^2 2s^2 2p^6 3s^2 3p^3$

99. The electron configuration of bromine shows that it is one electron short of having the same electron configuration of argon, a noble gas. Therefore bromine is very reactive because it wants to obtain a full outer shell of electrons as the noble gases already have. Krypton is a noble gas that already has a full outer shell, therefore there is no advantage to undergoing any reactions.

101. Oxidation is the loss of an electron and is related to ionization energy. The trend for ionization energy is that it decreases as you move downward and toward the left side of the periodic table. Therefore, K is the most easily oxidized.

103. $$E=\frac{hc}{\lambda} \Rightarrow \lambda=\frac{hc}{E}=\frac{(6.626 \otimes 10^{-34}\,J \cdot s)(3.00 \times 10^{8}\,m/s)}{3.0 \times 10^{-19}\,J}=6.6 \times 10^{-7}\,m$$

105. $$1.496 \times 10^{8}\,km \times \frac{1 \times 10^{3}\,m}{1\,km} \times \frac{1\,s}{3.00 \times 10^{8}\,m} = 498.7\,s = 8.31\,min = 8\,min\ 19\,sec$$

107. The quantum-mechanical model provided the ability to understand and predict chemical bonding, which is the basic level of understanding of matter and how it interacts. This model was critical in the areas of lasers, computers, drug design and semiconductors. The quantum-mechanical model for the atom is considered the foundation of modern chemistry.

109. a) $\lambda = \frac{h}{mv} = \frac{(6.626\times10^{-34}\,\text{J}\cdot\text{s})}{4.59\times10^{-5}\,\text{kg}\times95\,m/s} = 1.5\times10^{-31}\,\text{m}$

b) $\lambda = \frac{h}{mv} = \frac{(6.626\times10^{-34}\,\text{J}\cdot\text{s})}{9.109\times10^{-31}\,\text{kg}\times3.88\times10^{6}\,m/s} = 1.87\times10^{-10}\,\text{m}$

The wavelength of an e^- is a measurable value. The wavelength of a golf-ball is so small that it is not a practical value to even attempt to measure.

111. The dips in ionization energy occur when the electron being removed occupies a more stable location than the following electron of the following element. This occurs for the full *s* orbital (2A) and when the *p* orbital is half full (5A), illustrating the stability associated with full and half-full orbitals.

Highlight Problems

113. a) $E = \frac{hc}{\lambda} = \frac{(6.626\times10^{-34}\,\text{J}\cdot\text{s})(3.00\times10^{8}\,\text{m/s})}{1500\,\text{nm}\times\frac{1\,\text{m}}{1\times10^{9}\,\text{nm}}} = 1.3\times10^{-19}\,\text{J per photon}$

$\frac{1.3\times10^{-19}\,\text{J}}{\text{photon}}\times\frac{1\,\text{kJ}}{1\times10^{3}\,\text{J}}\times\frac{6.022\times10^{23}\,\text{photons}}{1\,\text{mole}} = 8.0\times10^{1}\,\text{kJ/mol}$, No.

b) $E = \frac{hc}{\lambda} = \frac{(6.626\times10^{-34}\,\text{J}\cdot\text{s})(3.00\times10^{8}\,\text{m/s})}{500\,\text{nm}\times\frac{1\,\text{m}}{1\times10^{9}\,\text{nm}}} = 3.98\times10^{-19}\,\text{J per photon}$

$\frac{3.98\times10^{-19}\,\text{J}}{\text{photon}}\times\frac{1\,\text{kJ}}{1\times10^{3}\,\text{J}}\times\frac{6.022\times10^{23}\,\text{photons}}{1\,\text{mole}} = 239\,\text{kJ/mol}$, No.

c) $E = \frac{hc}{\lambda} = \frac{(6.626\times10^{-34}\,\text{J}\cdot\text{s})(3.00\times10^{8}\,\text{m/s})}{150\,\text{nm}\times\frac{1\,\text{m}}{1\times10^{9}\,\text{nm}}} = 1.3\times10^{-18}\,\text{J per photon}$

$\frac{1.3\times10^{-18}\,\text{J}}{\text{photon}}\times\frac{1\,\text{kJ}}{1\times10^{3}\,\text{J}}\times\frac{6.022\times10^{23}\,\text{photons}}{1\,\text{mole}} = 800\,\text{kJ/mol}$, Yes.

Chemical Bonding

Questions

1. Bonding theories are important because they predict how atoms bond together to form compounds. They predict what combination of atoms form compounds and what combinations do not. Bonding theories can explain the shapes of molecules which in turn determine many of their physical and chemical properties.

3. An octet is an atom that contains eight valence electrons. A duet is a pair of valence electrons. A chemical bond is the sharing or transfer of electrons to form a stable electron configuration.

5. The Lewis structure for potassium has one valence electron, while the Lewis Structure for monoatomic chlorine has seven valence electrons. From these structures we can determine that if potassium gives up the one valence electron to chlorine, K^+ and Cl^- are formed. Therefore, the formula must be KCl.

7. A double bond is shorter and stronger than a single bond. A triple bond is shorter and even stronger than a double bond or a single bond.

9. Add up the valence electrons from each atom that is forming the molecule.

11. The octet rule has exceptions because the theory is not 100% accurate; however, it does work well in a majority of cases. Some exceptions to the rules are compounds that have odd numbers of valence electrons, boron compounds that tend to form with only six valence electrons, and some compounds that have more than eight valence electrons.

13. The valence shell electron pair repulsion (VSEPR) theory states that molecular shape is dictated by the fact that electron pairs, whether lone pairs or bonding pairs, repel each other and try to assume an optimum maximum distance from each other.

15. a) 180°
    b) 120°
    c) 109.5°

17. Electronegativity is the ability of an atom to attract electrons toward itself in a covalent bond.

19. A polar covalent bond is a covalent bond in which electrons are not equally shared between the two atoms.

21. When you try to mix a polar liquid with a nonpolar liquid they will separate to form two distinct regions.

Problems

Writing Lewis Structures for Elements

23. a) $1s^2\underline{2s^22p^3}$, $\cdot\dot{\underset{\cdot}{\text{N}}}:$

    b) $1s^2\underline{2s^22p^2}$, $\cdot\dot{\underset{\cdot}{\text{C}}}\cdot$

    c) $1s^22s^22p^6\underline{3s^23p^5}$, $:\ddot{\underset{\cdot\cdot}{\text{Cl}}}\cdot$

    d) $1s^22s^22p^6\underline{3s^23p^6}$, $:\ddot{\underset{\cdot\cdot}{\text{Ar}}}:$

25. a) $:\ddot{\underset{\cdot\cdot}{\text{I}}}\cdot$

    b) $:\dot{\underset{\cdot\cdot}{\text{S}}}\cdot$

    c) $\cdot\dot{\underset{\cdot}{\text{Ge}}}\cdot$

    d) $\cdot\dot{\text{Ca}}$

27. $:\ddot{\underset{\cdot\cdot}{\text{X}}}:$ Halogens tend to gain one electron in reactions.

29. M: Alkaline earth metals tend to lose two electrons in reactions.

31. a) Al^{3+}
    b) Mg^{2+}
    c) $\left[:\ddot{\underset{\cdot\cdot}{\text{Se}}}:\right]^{2-}$
    d) $\left[:\ddot{\underset{\cdot\cdot}{\text{N}}}:\right]^{3-}$

33. a) Kr
    b) Ne
    c) Kr
    d) Xe

Lewis Structures for Ionic Compounds

35. a) covalent
    b) ionic
    c) covalent
    d) ionic

37. a) $Na^{+}\ [:\ddot{\underset{\cdot\cdot}{F}}:]^{-}$
    b) $Ca^{2+}\ [:\ddot{\underset{\cdot\cdot}{O}}:]^{2-}$
    c) $[:\ddot{\underset{\cdot\cdot}{Br}}:]^{-}\ Sr^{2+}\ [:\ddot{\underset{\cdot\cdot}{Br}}:]^{-}$
    d) $K^{+}\ [:\ddot{\underset{\cdot\cdot}{O}}:]^{2-}\ K^{+}$

39. a) CaS
    b) $MgBr_2$
    c) CsI
    d) Ca_3N_2

41. a) $[:\ddot{\underset{\cdot\cdot}{F}}:]^{-}\ Mg^{2+}[:\ddot{\underset{\cdot\cdot}{F}}:]^{-}$
    b) $Mg^{2+}\ [:\ddot{\underset{\cdot\cdot}{O}}:]^{2-}$
    c) $Mg^{2+}[:\ddot{\underset{\cdot\cdot}{N}}:]^{3-}Mg^{2+}[:\ddot{\underset{\cdot\cdot}{N}}:]^{3-}Mg^{2+}$

43. a) $Cs^{+}\ [:\ddot{\underset{\cdot\cdot}{Cl}}:]^{-}$
    b) $Ba^{2+}\ [:\ddot{\underset{\cdot\cdot}{O}}:]^{2-}$
    c) $[:\ddot{\underset{\cdot\cdot}{I}}:]^{-}\ Ca^{2+}\ [:\ddot{\underset{\cdot\cdot}{I}}:]^{-}$

Lewis Structures for Covalent Compounds

45. a) A single hydrogen atom has one valence electron. When two hydrogen atoms share a single valence electron with the other, they each get a duet, a stable configuration for hydrogen.
    b) A single iodine atom has seven valence electrons. When two iodine atoms share a single valence electron with the other, they each get an octet.
    c) A single nitrogen atom has five valence electrons. When two nitrogen atoms share three valence electrons with the other, they each get an octet.
    d) A single oxygen atom has six valence electrons. When two oxygen atoms share a pair of electrons with the other, they each get an octet.

47. a) H— P —H
　　　　|
　　　H

b) :Cl—S—Cl:

c) :F—F:

d) H—I:

49. a) O=O

b) :C ≡ O:

c) H—O—N=O or H—O=N—O:

d) O=S—O: ↔ :O—S=O

51. a) H—C≡C—H

b) H— C = C —H
　　　|　 |
　　　H　H

c) H—N=N—H

d) H— N — N —H
　　　|　　|
　　　H　 H

53. a) :N ≡ N:

b) S=Si=S

c) H—O—H

d) :I— N —I:
　　　|
　　 :I:

55. a) O=Se—O: ↔ :O—Se=O

b) $\left[\text{O} = \text{C} - \text{:O:} \right]^{2-}$ (with :O: bonded below C) ↔ $\left[\text{:O} - \text{C} = \text{O} \right]^{2-}$ (with :O: bonded below C) ↔ $\left[\text{:O} - \text{C} - \text{O:} \right]^{2-}$ (with :O: double-bonded below C)

c) $[\text{:Cl—O:}]^{-}$

d) $[\text{:O—Cl—O:}]^{-}$

57. a) $\left[\begin{array}{c} :\ddot{O}: \\ | \\ :\underset{..}{\ddot{O}}-P-\underset{..}{\ddot{O}}: \\ | \\ :\underset{..}{O}: \end{array}\right]^{3-}$

b) $[:C \equiv N:]^{-}$

c) $[\underset{..}{\ddot{O}}{=}\dot{N}{-}\underset{..}{\ddot{O}}:]^{-} \leftrightarrow [:\underset{..}{\ddot{O}}{-}\dot{N}{=}\underset{..}{\ddot{O}}]^{-}$

d) $\left[\begin{array}{c} :\underset{..}{\ddot{O}}-\underset{..}{\ddot{S}}-\underset{..}{\ddot{O}}: \\ | \\ :\underset{..}{O}: \end{array}\right]^{2-}$

59. a) $\begin{array}{c} :\underset{..}{\ddot{Cl}}-\underset{}{B}-\underset{..}{\ddot{Cl}}: \\ | \\ :\underset{..}{Cl}: \end{array}$

b) $\underset{..}{\ddot{O}}{=}\dot{N}{-}\underset{..}{\ddot{O}}: \leftrightarrow :\underset{..}{\ddot{O}}{-}\dot{N}{=}\underset{..}{\ddot{O}}$

c) $\begin{array}{c} H-B-H \\ | \\ H \end{array}$

Predicting the Shapes of Molecules

61. a) 4
    b) 4
    c) 2
    d) 4

63. a) 2 bonding groups, 2 lone pairs
    b) 3 bonding groups, 1 lone pair
    c) 2 bonding groups, 0 lone pair
    d) 4 bonding groups, 0 lone pair

65. a) tetrahedral
    b) trigonal planar
    c) linear
    d) trigonal planar

67. a) 109.5°
    b) 120°
    c) 180°
    d) 120°

69. a) electron geometry = linear
   molecular geometry = linear
   b) electron geometry = trigonal planar
   molecular geometry = bent
   c) electron geometry = tetrahedral
   molecular geometry = bent
   d) electron geometry = tetrahedral
   molecular geometry = trigonal pyramidal

71. a) 180°
   b) 120°
   c) 109.5°
   d) 109.5°

73. a) electron geometry = linear
   molecular geometry = linear
   b) Both nitrogen atoms have identical electron and molecular geometry.
   electron geometry = trigonal planar
   molecular geometry = bent
   c) Both nitrogen atoms have identical electron and molecular geometry.
   electron geometry = tetrahedral
   molecular geometry = trigonal pyramidal

75. a) trigonal planar
   b) bent
   c) trigonal planar
   d) tetrahedral

Electronegativity and Polarity

77. a) 1.2
   b) 1.8
   c) 2.8

79. Cl>Si>Ga>Ca>Rb

81. a) 2.8 − 1.2 = 1.6, polar covalent
   b) 4.0 − 1.6 = 2.4, ionic
   c) 2.8 − 2.8 = 0, pure covalent
   d) 3.5 − 1.8 = 1.7, polar covalent

83.

Atom 1 (E.N.)	Atom 2 (E.N.)	E.N. Difference
H (2.1)	H (2.1)	0.0
I (2.5)	Cl (3.0)	0.5
H (2.1)	Br (2.8)	0.7
C (2.5)	O (3.5)	1.0

Trend: $H_2 < ICl < HBr < CO$

85. a) polar, different electronegativities
    b) nonpolar, same electronegativities
    c) nonpolar, same electronegativities
    d) polar, different electronegativities

87. a) :C≡O:
    ⟼
    b) nonpolar
    c) nonpolar
    d) H–B̈r:
    ⟼

89. a) nonpolar, The C and S atoms have identical electronegativities.
    b) polar
    c) nonpolar, The C and H atoms have different electronegativities, but symmetry cancels the effect.
    d) polar

91. a) nonpolar, The B and H atoms have nearly identical electronegativities, so the bonds are not sufficiently polar and the symmetry of the molecule would cancel out any polarity effects.
    b) polar
    c) nonpolar, The C and H atoms have a small difference in electronegativities (0.4), and symmetry cancels out any polarity effects.
    d) polar

Cumulative Problems

93. a) Ca: $1s^2 2s^2 2p^6 3s^2 3p^6 \underline{4s^2}$, ·Ca·
    b) Ga: $1s^2 2s^2 2p^6 3s^2 3p^6 \underline{4s^2} 3d^{10} \underline{4p^1}$, ·Ġa·
    c) As: $1s^2 2s^2 2p^6 3s^2 3p^6 \underline{4s^2} 3d^{10} \underline{4p^3}$, ·Ȧs:
    d) I: $1s^2 2s^2 2p^6 3s^2 3p^6 4s^2 3d^{10} 4p^6 \underline{5s^2} 4d^{10} \underline{5p^5}$, :Ï:

95. a) ionic, $K^+ [:\ddot{S}:]^{2-} K^+$

b) covalent, H— C =Ö (with :F: bonded to C)

c) ionic, $Mg^{2+} [:\ddot{Se}:]^{2-}$

d) covalent, :Br— P —Br (with :Br: bonded to P)

97. :Cl—C(=O:)—Cl: ; :Cl—C(=O:)—Cl:

Polar Trigonal Planar

99. H—C(H)(H)—C(=O:)—ÖH ; H—C(H)(H)—C(=O)—OH

Polar

101. $H:\ddot{Cl}: + Na^+[:\ddot{O}–H]^- \rightarrow H–\ddot{O}–H + Na^+[:\ddot{Cl}:]^-$

103. $K\cdot$, $:\ddot{Cl}–\ddot{Cl}: \Rightarrow [K]^+[:\ddot{Cl}:]^-$, K was oxidized and Cl_2 was reduced

105. a) $K^+[:\ddot{O}–H]^-$

b) $K^+[:\ddot{O}–N(–:\ddot{O}:)=\ddot{O}]^- \leftrightarrow K^+[\ddot{O}=N(–:\ddot{O}:)–\ddot{O}:]^- \leftrightarrow K^+[:\ddot{O}–N(=\ddot{O}:)–\ddot{O}:]^-$

c) $Li^+[:\ddot{I}–\ddot{O}:]^-$

d) $Ba^{2+}[:\ddot{O}–C(–:\ddot{O}:)=\ddot{O}]^{2-} \leftrightarrow Ba^{2+}[\ddot{O}=C(–:\ddot{O}:)–\ddot{O}:]^{2-} \leftrightarrow Ba^{2+}[:\ddot{O}–C(=\ddot{O}:)–\ddot{O}:]^{2-}$

107.

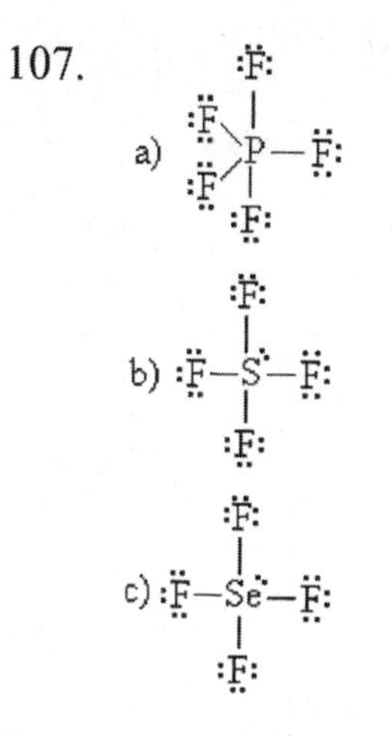

109. $\frac{46.02\text{g}}{1\text{mol}} \times \frac{26.10\%\text{C}}{100} = 12.01\text{g} \Rightarrow 1\text{mol C}$

$\frac{46.02\text{g}}{1\text{mol}} \times \frac{4.38\%\text{H}}{100} = 2.02\text{g} \Rightarrow 2\text{mol H}$

$\frac{46.02\text{g}}{1\text{mol}} \times \frac{69.52\%\text{O}}{100} = 32.00\text{g} \Rightarrow 2\text{mol O}$

Formula=CH_2O_2

```
  :O:
   ||  ..
H-C-O-H
       ..
```

111. The best possible structure for HOO is H-Ö-Ö· ;this is not a stable molecule as the second oxygen atom does not have an octet. The electron geometry is tetrahedral and a bent molecular geometry.

Highlight Problems

113. a) $[:\ddot{O}\text{-}\ddot{O}:]^-$

b) $[:\ddot{O}:]^-$

c) :Ö–H

d)
```
    H
    |
H–C–Ö–Ö·
    |
    H
```

115. a) incorrect; The structure should have tetrahedral electron geometry and a bent molecular geometry.

H–S̈e–H

b) correct

c) incorrect; The structure should have tetrahedral electron geometry and a trigonal pyramidal geometry.

```
:C̈l–P̈–C̈l:
     |
    :C̈l:
```

d) correct

Gases

11

Questions

1. Pressure is the force per unit area that is caused by gaseous molecules as they collide with a surface.

3. The kinetic molecular theory assumptions:
   - Gas particles are in constant motion and move in straight line.
   - Gas particles do not interact with each other (no attraction or repulsion). When particles collide, they bounce back like perfect billiard balls.
   - There is a large distance between gas particles compared to the size of each particle.
   - The average kinetic energy (energy of motion) is proportional to the temperature of the gas.

5. Ear pain experienced during changes in altitude is caused by air cavities within our ear. When the external pressure lowers, the air within the cavity is at a higher pressure. This causes the eardrum to bulge outward, which is the source of pain.

7. Boyle's law describes the relationship between pressure and volume of a gas. It states that pressure and volume are inversely related, or $V \propto 1/P$. As pressure increases, volume decreases (n and T being constant). When the volume of a gas sample is decreased, the same number of gas particles is crowded into a smaller volume, causing more collisions with the walls of the container and therefore increasing the pressure.

9. Extra long snorkels would not work because the pressure of the air on the surface is at about 1 atm of pressure while the air in a divers lung is at high pressure (10 m depth = 2 atm pressure). If you connected the diver with a snorkel at the surface, the high pressure in the lungs would force all of the air out to the surface, the opposite of what you want!

11. The gas molecules in a hot air balloon are moving faster and, according to Charles's law, will occupy a larger volume than the same amount of gas at ambient temperature. As the volume occupied by a given amount of gas increases, the density of the gas decreases and the balloon rises due to the buoyancy effect. The air inside the balloon is less dense than air outside the balloon.

13. Avogadro's law states that the volume of a gas and the amount of the gas are directly proportional. If the number of gas particles increases at constant pressure and temperature, the particles must occupy more volume.

15. The ideal gas law is most accurate under the conditions of low pressure and high temperature. It breaks down at high pressures and low temperatures. This breakdown occurs because the gases are no longer acting according to the kinetic molecular theory. The gas particles start interacting (attraction and repulsion), and the distance between gas particles is no longer large.

17. Dalton's law of partial pressure states that the total pressure exerted by a gas is equal to the sum of the partial pressures of the various component gases in the mixture: $P_{total} = P_1 + P_2 + P_3 + \ldots$

19. Deep-sea divers use a helium-oxygen mixture to prevent oxygen toxicity and nitrogen narcosis.

21. The vapor pressure of a liquid is equal to the equilibrium partial pressure of the gas in contact with the liquid. As temperature increases, the vapor pressure of a liquid increases.

Problems

Converting Between Pressure Units

23. a) $1277 \text{ mm Hg} \times \frac{1 \text{ atm}}{760 \text{ mm Hg}} = 1.680 \text{ atm}$

b) $2.38 \times 10^5 \text{Pa} \times \frac{1 \text{ atm}}{101{,}325 \text{ Pa}} = 2.35 \text{ atm}$

c) $127 \text{ psi} \times \frac{1 \text{ atm}}{14.7 \text{ psi}} = 8.64 \text{ atm}$

d) $455 \text{ torr} \times \frac{1 \text{ atm}}{760 \text{ torr}} = 0.599 \text{ atm}$

25. a) $2.3 \text{ atm} \times \frac{760 \text{ torr}}{1 \text{ atm}} = 1.7 \times 10^3 \text{ torr}$

b) $4.7x10^{-2} \text{atm} \times \frac{760 \text{ mm Hg}}{1 \text{ atm}} = 36 \text{ mm Hg}$

c) $24.8 \text{ psi} \times \frac{760 \text{ mm Hg}}{14.7 \text{ psi}} = 1.28 \times 10^3 \text{mm Hg}$

d) $32.84 \text{ in Hg} \times \frac{760 \text{ torr}}{29.92 \text{ in Hg}} = 834.2 \text{ torr}$

27.

Pascals	Atmospheres	mmHg	Torr	psi
882	0.00870	6.62	6.62	0.128
5.65×10^4	0.558	424	424	8.20
1.71×10^5	1.69	1.28×10^3	1.28×10^3	24.8
1.02×10^5	1.01	764	764	14.8
3.32×10^4	0.328	249	249	4.82

29. a) $24.9 \text{ in Hg} \times \frac{1 \text{ atm}}{29.92 \text{ in Hg}} = 0.832 \text{ atm}$

b) $24.9 \text{ in Hg} \times \frac{760 \text{ mm Hg}}{29.92 \text{ in Hg}} = 632 \text{ mm Hg}$

c) $24.9 \text{ in Hg} \times \frac{14.7 \text{ psi}}{29.92 \text{ in Hg}} = 12.2 \text{ psi}$

d) $24.9 \text{ in Hg} \times \frac{101{,}325 \text{ Pa}}{29.92 \text{ in Hg}} = 8.43\times10^4 \text{ Pa}$

31. a) $31.85 \text{ in Hg} \times \frac{760 \text{ mm Hg}}{29.92 \text{ in Hg}} = 809.0 \text{ mm Hg}$

b) $31.85 \text{ in Hg} \times \frac{1 \text{ atm}}{29.92 \text{ in Hg}} = 1.065 \text{ atm}$

c) $31.85 \text{ in Hg} \times \frac{760 \text{ torr}}{29.92 \text{ in Hg}} = 809.0 \text{ torr}$

d) $31.85 \text{ in Hg} \times \frac{101{,}325 \text{ Pa}}{29.92 \text{ in Hg}} \times \frac{1 \text{ kPa}}{1000 \text{ Pa}} = 107.9 \text{ kPa}$

Simple Gas Laws

33. $P_1 = 705 \text{ mm Hg},\ V_1 = 3.95 \text{ L},\ V_2 = 5.38 \text{ L},\ P_2 = ?;\ P_1V_1 = P_2V_2$

$(705 \text{ mm Hg})(3.95 \text{ L}) = (P_2)(5.38 \text{ L})$

$$P_2 = \frac{(705 \text{ mm Hg})(3.95 \text{ L})}{(5.38 \text{ L})} = 518 \text{ mm Hg}$$

35. $P_1 = 1.0 \text{ atm},\ V_1 = 6.3 \text{ L},\ V_2 = ? \text{ L},\ P_2 = 3.5 \text{ atm};\ P_1V_1 = P_2V_2$

$$(1.0 \text{ atm})(6.3 \text{ L}) = (V_2)(3.5 \text{ atm}) \Rightarrow V_2 = \frac{(1.0 \text{ atm})(6.3 \text{ L})}{(3.5 \text{ atm})} = 1.8 \text{ L}$$

37.

P1	V1	P2	V2
755 mm Hg	2.85 L	885 mmHg	2.43 L
9.35 atm	1.33 L	4.32 atm	2.88 L
192 mm Hg	382 mL	152 mmHg	482 mL
2.11 atm	226 mL	3.82 atm	125 mL

39. T_1=299 K, V_1=3.2 L, T_2=376 K, V_2=?; $\frac{V_1}{T_1}=\frac{V_2}{T_2}$

$$\frac{3.2\text{ L}}{299\text{ K}}=\frac{V_2}{376\text{ K}} \Rightarrow V_2=\frac{(3.2\text{ L})(376\text{ K})}{(299\text{ K})}=4.0\text{ L}$$

41. T_1=22+273=295 K, V_1=48.3 mL, T_2=87+273=3.60×10^2 K, V_2=?; $\frac{V_1}{T_1}=\frac{V_2}{T_2}$

$$\frac{48.3\text{ mL}}{295\text{ K}}=\frac{V_2}{3.60\times10^2\text{ K}} \Rightarrow V_2=\frac{(48.3\text{ mL})(3.60\times10^2\text{ K})}{(295\text{ K})}=58.9\text{ mL}$$

43.

V1	T1	V2	T2
1.08 L	25.4 ºC	1.33 L	94.5 ºC
58.9 mL	77 K	228 mL	298 K
115 cm^3	12.5 ºC	119 cm^3	22.4 ºC
232 L	18.5 ºC	294 L	96.2 ºC

45. n_1=0.12mol, V_1=2.55 L, n_2=0.32 mol, V_2=?; $\frac{V_1}{n_1}=\frac{V_2}{n_2}$

$$\frac{2.55\text{L}}{0.12\text{mol}}=\frac{V_2}{0.32\text{ mol}} \Rightarrow V_2=\frac{(2.55\text{L})(0.32\text{ mol})}{0.12\text{mol}}=6.8\text{ L}$$

47. n_1=0.128 mol, V_1=2.76 L, n_2=0.128+0.073=0.201 mol, V_2 = ?; $\frac{V_1}{n_1}=\frac{V_2}{n_2}$

$$\frac{2.76\text{ L}}{0.128\text{ mol}}=\frac{V_2}{0.201\text{ mol}} \Rightarrow V_2=\frac{(2.76\text{ L})(0.201\text{ mol})}{(0.128\text{ mol})}=4.33\text{ L}$$

49.

V1	n1	V2	n2
38.5 mL	$1.55\text{x}10^{-3}$ mol	49.4 mL	$1.99\text{x}10^{-3}$ mol
8.03 L	1.37 mol	26.8 L	4.57 mol
11.2 L	0.628 mol	15.7 L	0.881 mol
422 mL	0.0109 mol	671 mL	0.0174 mol

Combined Gas Law

51. $P_1 = 725$ mm Hg, $V_1 = 28.4$ L, $T_1 = 305$ K, $V_2 = 14.8$ L, $T_2 = 375$ K, $P_2 = ?$

$$\frac{P_1V_1}{T_1} = \frac{P_2V_2}{T_2} \Rightarrow \frac{(725 \text{ mm Hg})(28.4 \text{ L})}{(305 \text{ K})} = \frac{P_2(14.8 \text{ L})}{(375 \text{ K})}$$

$$P_2 = \frac{(725 \text{ mm Hg})(28.4 \text{ L})(375 \text{ K})}{(305 \text{ K})(14.8 \text{ L})} = 1.71 \times 10^3 \text{ mm Hg}$$

53. P_1=1.0 atm, V_1=2.8 L, T_1=34+273=307 K,
P_2=3.5 atm, $V_2 = ?$, $T_2 = 18 + 273 = 291$ K

$$\frac{P_1V_1}{T_1} = \frac{P_2V_2}{T_2} \Rightarrow \frac{(1.0 \text{ atm})\ (2.8 \text{ L})}{(307 \text{ K})} = \frac{(3.5 \text{ atm})(V_2)}{(291 \text{ K})} \Rightarrow$$

$$V_2 = \frac{(1.0 \text{ atm})(2.8 \text{ L})(291 \text{ K})}{(307 \text{ K})(3.5 \text{ atm})} = 0.76 \text{ L}$$

55. P_1=735 mm Hg, T_1=28+273=301 K, P_2=?, T_2=86+273=359 K

$$\frac{P_1}{T_1} = \frac{P_2}{T_2} \Rightarrow \frac{(735 \text{ mm Hg})}{(301 \text{ K})} = \frac{(P_2)}{(359 \text{ K})} \Rightarrow P_2 = \frac{(735 \text{ mm Hg})(359 \text{ K})}{(301 \text{ K})} = 877\text{mm Hg}$$

57.

P1	V1	T1	P2	V2	T2
121 atm	1.58 L	12.2 °C	1.54 atm	133 L	32.3 °C
721 torr	141 mL	135 K	801 torr	152 mL	162 K
5.51 atm	0.879 L	22.1 °C	4.86 atm	1.05 L	38.3 °C

The Ideal Gas Las

59. $P = 1.25$ atm, $V = ?$, $n = 0.255$ mol, $T = 305$ K, $PV = nRT$

$$(1.25 \text{ atm})\ V = (0.255 \text{ mol})(0.0821 \frac{\text{L} \cdot \text{atm}}{\text{mol} \cdot \text{K}})(305 \text{ K})$$

$$V = \frac{(0.255 \text{ mol})(0.0821 \frac{\text{L} \cdot \text{atm}}{\text{mol} \cdot \text{K}})(305 \text{ K})}{(1.25 \text{ atm})} = 5.11 \text{ L}$$

61. P=1.8 atm, V=28.5 L, n=?, T=298 K, PV=nRT

$$(1.8 \text{ atm})(28.5 \text{ L})=(n)\ (0.0821 \frac{\text{L} \cdot \text{atm}}{\text{mol} \cdot \text{K}})(298 \text{ K}) \Rightarrow$$

$$n= \frac{(1.8 \text{ atm})(28.5 \text{ L})}{(0.0821 \frac{\text{L} \cdot \text{atm}}{\text{mol} \cdot \text{K}})(298 \text{ K})} = 2.1 \text{ mol}$$

63. $P = 43.2 \text{ psi} \times \frac{1 \text{ atm}}{14.7 \text{ psi}} = 2.94 \text{ atm}, V = 11.8 \text{ L}, n = ?, T = 25+273 = 298 \text{ K}$

$$PV = nRT \Rightarrow (2.94 \text{ atm})(11.8 \text{ L}) = (n)(0.0821 \frac{\text{L} \cdot \text{atm}}{\text{mol} \cdot \text{K}})(298 \text{ K}) \Rightarrow$$

$$n = \frac{(2.94 \text{ atm})(11.8 \text{ L})}{(0.0821 \frac{\text{L} \cdot \text{atm}}{\text{mol} \cdot \text{K}})(298 \text{ K})} = 1.42 \text{ mol}$$

65.

P	V	n	T
1.05 atm	1.19 L	0.112 mol	136 K
112 torr	40.9 L	0.241 mol	304 K
1.49 atm	28.5 mL	1.74×10^{-3} mol	25.4 °C
0.559 atm	0.439 L	0.0117 mol	255 K

67. $P = 24.1 \text{ psi} \times \frac{1 \text{ atm}}{14.7 \text{ psi}} = 1.64 \text{ atm}, V = 3.5 \text{ L}, n = ?, T = 25+273 = 298 \text{ K}$

$$PV = nRT \Rightarrow (1.64 \text{ atm})(3.5 \text{ L}) = (n)(0.0821 \frac{\text{L} \cdot \text{atm}}{\text{mol} \cdot \text{K}})(298 \text{ K}) \Rightarrow$$

$$n = \frac{(1.64 \text{ atm})(3.5 \text{ L})}{(0.0821 \frac{\text{L} \cdot \text{atm}}{\text{mol} \cdot \text{K}})(298 \text{ K})} = 0.23 \text{ mol}$$

69. $P = 745 \text{ mm Hg} \times \frac{1 \text{ atm}}{760 \text{ mm Hg}} = 0.980 \text{ atm}, V = 248 \text{ mL} \times \frac{1 \text{ L}}{1000 \text{ mL}} = 0.248 \text{ L},$
$n = ?, T = 28+273 = 301 \text{ K}$

$$PV = nRT \Rightarrow (0.980 \text{ atm})(0.248 \text{ L}) = (n)(0.0821 \frac{\text{L} \cdot \text{atm}}{\text{mol} \cdot \text{K}})(301 \text{ K}) \Rightarrow$$

$$n = \frac{(0.980 \text{ atm})(0.248 \text{ L})}{(0.0821 \frac{\text{L} \cdot \text{atm}}{\text{mol} \cdot \text{K}})(301 \text{ K})} = 0.00983 \text{ mol}$$

$$\text{molar mass} = \frac{0.433 \text{ g}}{0.00983 \text{ mol}} = 44.0 \text{ g/mol}$$

71. $P = 886 \text{ torr} \times \frac{1 \text{ atm}}{760 \text{ torr}} = 1.17 \text{ atm}$, $V = 224 \text{ mL} \times \frac{1 \text{ L}}{1000 \text{ mL}} = 0.224 \text{ L}$,

$n = ?$, $T = 55+273 = 328 \text{ K}$

$$PV = nRT \Rightarrow (1.17 \text{ atm})(0.224 \text{ L}) = (n)(0.0821 \frac{\text{L} \cdot \text{atm}}{\text{mol} \cdot \text{K}})(328 \text{ K}) \Rightarrow$$

$$n = \frac{(1.17 \text{ atm})(0.224 \text{ L})}{(0.0821 \frac{\text{L} \cdot \text{atm}}{\text{mol} \cdot \text{K}})(328 \text{ K})} = 0.00973 \text{ mol}$$

$$\text{molar mass} = \frac{38.8 \text{ mg} \times \frac{1 \text{ g}}{1000 \text{ mg}}}{0.00973 \text{ mol}} = 3.99 \text{ g/mol}$$

Partial Pressure

73. $P_{tot} = P_1 + P_2 + P_3 + \ldots \Rightarrow P_{tot} = 217 \text{ torr} + 106 \text{ torr} + 248 \text{ torr} = 571 \text{ torr}$

75. $P_{tot} = P_{He} + P_{O_2} \Rightarrow 11.0 \text{ atm} = P_{He} + 0.30 \text{ atm} \Rightarrow P_{He} = 11.0 - 0.30 = 10.7 \text{ atm}$

77. $P_{tot} = P_{H_2} + P_{H_2O} \Rightarrow 732 \text{ mm Hg} = P_{H_2} + 31.8$

$P_{H_2} = 732 - 31.8 = 7.00 \times 10^2 \text{ mm Hg}$

79. $P_{N_2} = 1.12 \text{ atm} \times 0.78 = 0.87 \text{ atm}$ $\quad P_{O_2} = 1.12 \text{ atm} \times 0.22 = 0.25 \text{ atm}$

81. $P_{O_2} = 8.5 \text{ atm} \times 0.040 = 0.34 \text{ atm}$

Molar Volume

83. a) $22.5 \text{ mol } Cl_2 \times \frac{22.4 \text{ L}}{1 \text{ mol } Cl_2} = 504 \text{ L}$

b) $3.6 \text{ mol } N_2 \times \frac{22.4 \text{ L}}{1 \text{ mol } N_2} = 81 \text{ L}$

c) $2.2 \text{ mol He} \times \frac{22.4 \text{ L}}{1 \text{ mol He}} = 49 \text{ L}$

d) $27 \text{ mol } CH_4 \times \frac{22.4 \text{ L}}{1 \text{ mol } CH_4} = 6.0 \times 10^2 \text{ L}$

85. a) $73.9\text{ g N}_2 \times \frac{1\text{ mol N}_2}{28.02\text{ g}} \times \frac{22.4\text{ L}}{1\text{ mol N}_2} = 59.1\text{ L}$

b) $42.9\text{ g O}_2 \times \frac{1\text{ mol O}_2}{32.00\text{ g}} \times \frac{22.4\text{ L}}{1\text{ mol O}_2} = 30.0\text{ L}$

c) $148\text{ g NO}_2 \times \frac{1\text{ mol NO}_2}{46.01\text{ g}} \times \frac{22.4\text{ L}}{1\text{ mol NO}_2} = 72.1\text{ L}$

d) $245\text{ mg CO}_2 \times \frac{1\text{ g}}{1000\text{ mg}} \times \frac{1\text{ mol CO}_2}{44.01\text{ g}} \times \frac{22.4\text{ L}}{1\text{ mol CO}_2} = 0.125\text{ L}$

87. a) $178\text{ mL CO}_2 \times \frac{1\text{ L}}{1000\text{ mL}} \times \frac{1\text{ mol CO}_2}{22.4\text{ L}} \times \frac{44.01\text{ g}}{1\text{ mol CO}_2} = 0.350\text{ g}$

b) $155\text{ mL O}_2 \times \frac{1\text{ L}}{1000\text{ mL}} \times \frac{1\text{ mol O}_2}{22.4\text{ L}} \times \frac{32.00\text{ g}}{1\text{ mol O}_2} = 0.221\text{ g}$

c) $1.25\text{ L SF}_6 \times \frac{1\text{ mol SF}_6}{22.4\text{ L}} \times \frac{146.07\text{ g}}{1\text{ mol SF}_6} = 8.15\text{ g}$

Gases in Chemical Reactions

89. $n = 1.07\text{ mol C} \times \frac{1\text{ mol H}_2}{1\text{ mol C}} = 1.07\text{ mol H}_2,\ P = 1.0\text{ atm},\ T = 315\text{ K},\ V = ?$

$PV = nRT \Rightarrow (1.0\text{ atm})(V) = (1.07\text{ mol})(0.0821\frac{\text{L}\cdot\text{atm}}{\text{mol}\cdot\text{K}})(315\text{ K}) \Rightarrow$

$$V = \frac{(1.07\text{ mol})(0.0821\frac{\text{L}\cdot\text{atm}}{\text{mol}\cdot\text{K}})(315\text{ K})}{(1.0\text{ atm})} = 28\text{ L}$$

91. $P = 748\text{ mm Hg} \times \frac{1\text{ atm}}{760\text{ mm Hg}} = 0.984\text{ atm},\ V = ?,\ T = 86+273 = 359\text{ K}$

$n = 0.55\text{ mol CH}_3\text{OH} \times \frac{2\text{ mol H}_2}{1\text{ mol CH}_3\text{OH}} = 1.1\text{ mol H}_2$

$PV = nRT \Rightarrow (0.984\text{ atm})(V) = (1.1\text{ mol})(0.0821\frac{\text{L}\cdot\text{atm}}{\text{mol}\cdot\text{K}})(359\text{ K}) \Rightarrow$

$$V = \frac{(1.1\text{ mol})(0.0821\frac{\text{L}\cdot\text{atm}}{\text{mol}\cdot\text{K}})(359\text{ K})}{(0.984\text{ atm})} = 33\text{ L of H}_2$$

The molar ratio of CO to H_2 is 1:2, we can therefore divide the volume of H_2 in half to determine CO. Volume of CO = 33/2 = 17 L

**Warning*: this only works for gases! It does not apply to liquids or solids.*

93. $P = 892 \text{ torr} \times \frac{1 \text{ atm}}{760 \text{ torr}} = 1.17 \text{ atm}, V = ?, T = 95 + 273 = 368 \text{ K}$

$n = 18.5 \text{ g Al} \times \frac{1 \text{ mol Al}}{26.98 \text{ g}} \times \frac{1 \text{ mol } N_2}{2 \text{ mol Al}} = 0.343 \text{ mol } N_2$

$PV = nRT \Rightarrow (1.17 \text{ atm})(V) = (0.343 \text{ mol})(0.0821 \frac{\text{L} \cdot \text{atm}}{\text{mol} \cdot \text{K}})(368 \text{ K}) \Rightarrow$

$$V = \frac{(0.343 \text{ mol})(0.0821 \frac{\text{L} \cdot \text{atm}}{\text{mol} \cdot \text{K}})(368 \text{ K})}{(1.17 \text{ atm})} = 8.86 \text{ L}$$

95. $24.8 \text{ L } H_2 \times \frac{1 \text{ mol } H_2}{22.4 \text{ L}} \times \frac{2 \text{ mol } NH_3}{3 \text{ mol } H_2} \times \frac{17.04 \text{ g}}{1 \text{ mol } NH_3} = 12.6 \text{ g } NH_3$

97. $156.8 \text{ mL } O_2 \times \frac{1 \text{ L}}{1000 \text{ mL}} \times \frac{1 \text{ mol } O_2}{22.4 \text{ L}} \times \frac{2 \text{ mol Ca}}{1 \text{ mol } O_2} \times \frac{40.08 \text{ g}}{1 \text{ mol Ca}} = 0.561 \text{ g Ca}$

Cumulative Problems

99. $P = 1.00 \text{ atm}, V = ?, n = 1.00 \text{ mol}, T = 273 \text{ K}$

$$V = \frac{(1.00 \text{ mol})(0.0821 \frac{\text{L} \cdot \text{atm}}{\text{mol} \cdot \text{K}})(273 \text{ K})}{(1.00 \text{ atm})} = 22.4 \text{L}$$

101. $P = 267 \text{ torr} \times \frac{1 \text{ atm}}{760 \text{ torr}} = 0.351 \text{ atm}, V = 255 \text{ mL} \times \frac{1 \text{ L}}{1000 \text{ mL}} = 0.255 \text{ L},$
$n = ?, T = 25 + 273 = 298 \text{ K}, \text{mass} = 143.289 - 143.187 = 0.102 \text{ g}$

$PV = nRT \Rightarrow (0.351 \text{ atm})(0.255 \text{ L}) = (n)(0.0821 \frac{\text{L} \cdot \text{atm}}{\text{mol} \cdot \text{K}})(298\text{K})$

$$n = \frac{(0.351 \text{ atm})(0.255 \text{ L})}{(0.0821 \frac{\text{L} \cdot \text{atm}}{\text{mol} \cdot \text{K}})(298\text{K})} = 0.00366 \text{ moles}$$

$$\text{molar mass} = \frac{0.102 \text{ g}}{0.00366 \text{ moles}} = 27.9 \text{ g/mol}$$

103. $P = 556\text{ mm Hg} \times \frac{1\text{ atm}}{760\text{ mm Hg}} = 0.732\text{ atm}$, $V = 158\text{ mL} \times \frac{1\text{ L}}{1000\text{ mL}} = 0.158\text{ L}$,

$n = ?$, $T = 25+273 = 298\text{ K}$, mass = 0.275 g

$$PV = nRT \Rightarrow (0.732\text{ atm})(0.158\text{ L}) = (n)(0.0821\frac{\text{L}\cdot\text{atm}}{\text{mol}\cdot\text{K}})(298\text{K})$$

$$n = \frac{(0.732\text{ atm})(0.158\text{ L})}{(0.0821\frac{\text{L}\cdot\text{atm}}{\text{mol}\cdot\text{K}})(298\text{K})} = 0.00473\text{ moles}$$

$$\text{molar mass} = \frac{0.275\text{ g}}{0.00473\text{ moles}} = 58.1\text{ g/mol}$$

In one mole of compound,

$$\text{mass C} = 58.1 \times 0.8266 = 48.0\text{ g} \Rightarrow 48.0\text{ g} \times \frac{1\text{ mole C}}{12.01\text{ g}} = 4\text{ mol C}$$

$$\text{mass H} = 58.1 \times 0.1734 = 10.1\text{ g} \Rightarrow 10.1\text{ g} \times \frac{1\text{ mol H}}{1.01\text{ g}} = 10\text{ mol H}$$

Molecular formula: C_4H_{10}

105. $P_{tot} = P_{H_2O} + P_{H_2} \Rightarrow 748\text{ mm Hg} = 23.8 + P_{H_2} \Rightarrow P_{H_2} = 748 - 23.8 = 724\text{ mm Hg}$

$P = 724\text{ mm Hg} \times \frac{1\text{ atm}}{760\text{ mm Hg}} = 0.953\text{ atm}$, $T = 25+273 = 298\text{K}$, $n = ?$

$V = 325\text{ mL} \times \frac{1\text{ L}}{1000\text{ mL}} = 0.325\text{ L}$

$$PV = nRT \Rightarrow (0.953\text{ atm})(0.325\text{ L}) = (n)(0.0821\text{ L}\cdot\text{atm/mol}\cdot\text{K})(298\text{ K})$$

$$n = \frac{(0.953\text{ atm})(0.325\text{ L})}{(0.0821\text{L}\cdot\text{atm/mol}\cdot\text{K})(298\text{ K})} = 0.0127\text{ mol } H_2$$

$$0.0127\text{ mol } H_2 \times \frac{1\text{ mol Zn}}{1\text{ mol } H_2} \times \frac{65.39\text{ g}}{1\text{ mol Zn}} = 0.830\text{ g Zn}$$

107. $P_{tot} = P_{H_2O} + P_{H_2} \Rightarrow 748\text{ torr} = 55.3 + P_{H_2} \Rightarrow P_{H_2} = 748 - 55.3 = 693\text{ torr}$

$P = 693\text{ mm Hg} \times \frac{1\text{ atm}}{760\text{ mm Hg}} = 0.912\text{ atm}$, $T = 40+273 = 313\text{ K}$, $V = 1.78\text{ L}$, $n = ?$

$$PV = nRT \Rightarrow (0.912\text{ atm})(1.78\text{ L}) = (n)(0.0821\text{ L}\cdot\text{atm/mol}\cdot\text{K})(313\text{ K})$$

$$n = \frac{(0.912\text{ atm})(1.78\text{ L})}{(0.0821\text{ L}\cdot\text{atm/mol}\cdot\text{K})(313\text{ K})} = 0.0632\text{ mol } H_2$$

$$0.0632\text{ mol } H_2 \times \frac{2.02\text{ g}}{1\text{ mol } H_2} = 0.128\text{ g } H_2$$

109. $P_{tot} = P_{H_2O} + P_{O_2} \Rightarrow 752 \text{ mm Hg} = 23.8 + P_{O_2} \Rightarrow P_{O_2} = 752 - 23.8 = 728 \text{ mm Hg}$

$$P = 728 \text{ torr} \times \frac{1 \text{ atm}}{760 \text{ mm Hg}} = 0.958 \text{ atm, } T = 25{+}273 = 298 \text{ K, } V = ?$$

$$n = 15.8 \text{ g Ag} \times \frac{1 \text{ mol Ag}}{107.9 \text{ g}} \times \frac{1 \text{ mol } O_2}{4 \text{ mol Ag}} = 0.0366 \text{ mol}$$

$$PV = nRT \Rightarrow (0.958 \text{ atm})(V) = (0.0366)(0.0821 \text{ L}\cdot\text{atm/mol}\cdot\text{K})(298 \text{ K})$$

$$V = \frac{(0.0366)(0.0821 \text{ L}\cdot\text{atm/mol}\cdot\text{K})(298 \text{ K})}{(0.958 \text{ atm})} = 0.935 \text{ L}$$

111. $HCl(aq) + NaHCO_3(s) \rightarrow CO_2(g) + NaCl(aq) + H_2O(l)$

$$T = 22.7 + 273 = 296 \text{ K, } V = 28.2 \text{ mL} \times \frac{1 \text{ L}}{1000 \text{ mL}} = 0.0282 \text{ L}$$

$$PV = nRT \Rightarrow (0.954\text{atm})(0.0282\text{L}) = (n)(0.0821\frac{\text{L}\cdot\text{atm}}{\text{mol}\cdot\text{K}})(295.9 \text{ K}) \Rightarrow$$

$$n = \frac{(0.954\text{atm})(0.0282\text{L})}{(0.0821\frac{\text{L}\cdot\text{atm}}{\text{mol}\cdot\text{K}})(295.9 \text{ K})} = 1.11 \times 10^{-3} \text{mol } CO_2$$

$$1.11 \times 10^{-3} \text{mol } CO_2 \times \frac{1 \text{ mol } NaHCO_3}{1 \text{ mol } CO_2} \times \frac{84.01 \text{ g}}{1 \text{ mol } NaHCO_3} = 0.0933 \text{ g } NaHCO_3$$

113. a) $285.5 \text{ mL } SO_2 \times \frac{1 \text{ L}}{1000 \text{ mL}} \times \frac{1 \text{ mol } SO_2}{22.4 \text{ L}} \times \frac{2 \text{ mol } SO_3}{2 \text{ mol } SO_2} = 0.0127 \text{ mol } SO_3$

$$158.9 \text{ mL } O_2 \times \frac{1 \text{ L}}{1000 \text{ mL}} \times \frac{1 \text{ mol } O_2}{22.4 \text{ L}} \times \frac{2 \text{ mol } SO_3}{1 \text{ mol } O_2} = 0.0142 \text{ mol } SO_3$$

The limiting reactant is SO_2 and the theoretical yield is 0.0127 mol SO_3.

b) $187.2 \text{ mL } SO_3 \times \frac{1 \text{ L}}{1000 \text{ mL}} \times \frac{1 \text{ mol } SO_3}{22.4 \text{ L}} = 0.00836 \text{ mol } SO_3$

$$\text{Percent Yield} = \frac{\text{Actual}}{\text{Theoretical}} \times 100\% \Rightarrow \frac{0.00836}{0.0127} \times 100\% = 65.8\%$$

115. a) $12.8\text{ L NO}_2 \times \frac{1\text{ mol NO}_2}{22.4\text{ L}} \times \frac{2\text{ mol HNO}_3}{3\text{ mol NO}_2} = 0.381\text{ mol HNO}_3$

$14.9\text{ g H}_2\text{O} \times \frac{1\text{ mol H}_2\text{O}}{18.02\text{ g}} \times \frac{2\text{ mol HNO}_3}{1\text{ mol H}_2\text{O}} = 1.65\text{ mol HNO}_3$

Limiting reactant: NO_2, $\text{Mass} = 0.381\text{ mol HNO}_3 \times \frac{63.02\text{ g}}{1\text{ mol HNO}_3} = 24.0\text{ g HNO}_3$

b) $\text{Percent Yield} = \frac{\text{Actual}}{\text{Theoretical}} \times 100\% \Rightarrow \frac{14.8}{24.0} \times 100\% = 61.7\%$

117. $11.83\text{ g (NH}_4)_2\text{CO}_3 \times \frac{1\text{ mol (NH}_4)_2\text{CO}_3}{96.11\text{ g}} = 0.123\text{ mol (NH}_4)_2\text{CO}_3$

$0.123\text{ mol (NH}_4)_2\text{CO}_3 \times \frac{4\text{ mol gas}}{1\text{ mol (NH}_4)_2\text{CO}_3} = 0.492\text{ mol gas}$

$T = 22 + 273 = 295\text{ K}$

$PV = nRT \Rightarrow (1.02\text{ atm})(V) = (0.492\text{ mol})(0.0821\frac{\text{L}\cdot\text{atm}}{\text{mol}\cdot\text{K}})(295\text{ K}) \Rightarrow$

$$V = \frac{(0.492\text{ mol})(0.0821\frac{\text{L}\cdot\text{atm}}{\text{mol}\cdot\text{K}})(295\text{ K})}{(1.02\text{ atm})} = 11.7\text{ L}$$

119. $235\text{ mg He} \times \frac{1\text{ g}}{1000\text{ mg}} \times \frac{1\text{ mol He}}{4.00\text{ g}} = 0.0588\text{ mol He}$

$325\text{ mg Ne} \times \frac{1\text{ g}}{1000\text{ mg}} \times \frac{1\text{ mol Ne}}{20.18\text{ g}} = 0.0161\text{ mol Ne}$

$\text{Total Moles} = 0.0588 + 0.0161 = 0.0749\text{ mol}$

$P_{He} = \frac{0.0588\text{ mol He}}{0.0749\text{ mol Total}} \times 453\text{ torr} = 356\text{ torr}$

121. If the flask is filled with equal pressures of SO_2 and O_2 for a total pressure of 0.20 atm. According to the reaction two moles of SO_2 react with a single mole of O_2, therefore the SO_2 will react completely, leaving one-half of the O_2 unreacted (0.05 atm). The 0.10 atm of SO_2 that reacted forms an equal amount of SO_3 gas, therefore the final pressure of the vessel would be the unreacted O_2 and the newly formed SO_3 which is equal to 0.15 atm.

Highlight Problems

123. Choice c will have the greatest pressure because it has the highest number of gas particles that can collide with the container walls to create pressure.

125. $11.8\text{ L} \times \frac{1\text{ mol N}_2}{22.4\text{ L}} \times \frac{2\text{ mol NaN}_3}{3\text{ mol N}_2} \times \frac{65.02\text{ g}}{1\text{ mol NaN}_3} = 22.8\text{ g NaN}_3$

127. $\frac{V_1}{T_1} = \frac{V_2}{T_2} \Rightarrow \frac{(2.95\text{ L})}{(298\text{ K})} = \frac{(V_2)}{(77\text{ K})} \Rightarrow V_2 = \frac{(77\text{ K})(2.95\text{ L})}{(298\text{ K})} = 0.76\text{ L}$

The difference in the volumes (0.15L) is due to the fact that gas behavior is no longer ideal at extremely low temperatures.

Liquids, Solids, and Intermolecular Forces

Questions

1. Intermolecular forces are attractive forces that occur between individual molecules. Living organisms depend on intermolecular forces not only for taste but also for many other physiological processes. For example, intermolecular forces help determine the shapes of protein molecules and are central to DNA, the inheritable molecules that serve as blueprints for life.

3. The relative magnitude of the intermolecular forces to thermal energy determine whether a substance is a solid, liquid, or gas.

5. Properties of Solids
   - High densities in comparison to gases.
     -Solids have high densities in comparison to gases because the atoms or molecules that compose solids are also close together.
   - Definite shape; they do not assume the shape of their container.
     -The molecules or atoms that compose solids are fixed in place.
   - Definite volume; they are not easily compressed.
     -Solids have a definite volume and cannot be compressed because the molecules or atoms composing them are in close contact.
   - May be crystalline (ordered) or amorphous (disordered).
     -Solids may be crystalline, in which case the atoms or molecules that compose them arrange themselves in a well-ordered, three-dimensional array, or they may be amorphous, in which case the atoms or molecules that compose them have no long-range order.

7. Surface tension is the tendency of liquids to minimize their surface area due to the interaction of molecules between each other. The surface tension increases when intermolecular forces increase.

9. Evaporation is when a liquid undergoes a physical change to a gas. Condensation is the physical change when a gaseous substance changes to a liquid form.

11. The process of evaporation below the boiling point only occurs at the surface of the liquid. At the boiling point, there is sufficient thermal energy that molecules within the interior of the liquid can break free and enter into the gas phase.

13. The intermolecular forces in acetone are weaker, therefore it evaporates faster and is the more volatile compound.

15. Vapor pressure is the partial pressure of a gas in dynamic equilibrium with its liquid. The vapor pressure of a compound increases with increasing temperature and decreases with increasing strength of the intermolecular forces.

17. The process of condensation is the opposite of evaporation in that it is exothermic. This means that energy is released from the gas as it forms a liquid. Steam at 100°C releases more heat into your hand due to condensation than water at the same temperature.

19. The process of freezing is exothermic which releases heat from the water into the freezer. If the freezer cannot remove the excess heat, the freezer will start to warm up and the water will not freeze.

21. Melting ice is endothermic as the ice must absorb energy from the surroundings. The sign of ΔH is positive. Freezing water is an exothermic process as energy is released from the water into the surroundings and the sign of ΔH is negative.

23. Dispersion force (aka London force) is a type of intermolecular force present between all molecules and atoms. This type of intermolecular force arises from fluctuations in the spatial distribution of electrons in the atom or molecule that causes a temporary positive and negative charges within a particle. The dispersion force increases with increasing molar mass.

25. The hydrogen bond occurs when hydrogen bonds to either F, O, or N (very electronegative), which in turn pulls the shared electrons in the bond away from hydrogen. Hydrogen becomes very positive and forms a hydrogen bond to the lone pair electrons on F, O, or N of another molecule. Compounds that can form hydrogen bonds will have higher melting and boiling points than other compounds that do not form hydrogen bonds.

27. Molecular solids as a whole tend to have low to moderately low melting points relative to other types of solids; however, strong molecular forces can increase their melting points relative to each other.

29. Ionic solids tend to have much higher melting points relative to the melting points of other types of solids.

31. Water is unique because it has a low molar mass and is a liquid at room temperature, which is not the case with other low molar mass compounds. Additionally, while most compounds contract during the freezing process, water expands. This is the reason ice floats on water.

Problems

Evaporation, Condensation, Melting, and Freezing

33. The second beaker of 55 mL in a 12 cm diameter dish will evaporate more quickly because evaporation occurs at the surface and the 12 cm dish has a larger surface area.

35. Acetone will feel cooler because it has weaker intermolecular forces which make it more volatile. It will evaporate faster and, therefore, will remove more heat from your hand.

37. The temperature will increase from -5°C to the melting point at 0°C. The temperature will not increase until all of the ice has melted. After the ice has completely melted, the temperature will increase until it reaches room temperature (25°C).

39. The steam would cause a more severe burn because steam condensation is an exothermic process and this excess heat would increase the severity of the burn.

41. The water in the bag is undergoing the freezing process, which releases heat (exothermic) into the rest of the cooler.

43. The ice chest full of ice at 0°C would be colder because in order to melt the ice, heat from the cooler would be absorbed (endothermic). The ice chest full of water at 0°C would be warmer because the freezing process releases heat (exothermic) and would warm the cooler.

45. The boiling point occurs when the vapor pressure of the liquid is equal to the external pressure. Because of the altitude of Denver (1 mile above sea level), the atmospheric pressure is less than 1 atm. Therefore, the vapor pressure will match the atmospheric pressure at a lower temperature.

Heat of Vaporization and Heat of Fusion

47. $33.8 \text{ g } H_2O \times \frac{1 \text{ mol } H_2O}{18.02 \text{ g}} \times \frac{40.7 \text{ kJ}}{1 \text{ mol } H_2O} = 76.3 \text{ kJ}$

49. $2.8 \text{ g } H_2O \times \frac{1 \text{ mol } H_2O}{18.02 \text{ g}} \times \frac{40.7 \text{ kJ}}{1 \text{ mol } H_2O} = 6.3 \text{ kJ}$

51. $4.25 \text{ g } H_2O \times \frac{1 \text{ mol } H_2O}{18.02 \text{ g}} \times \frac{44.0 \text{ kJ}}{1 \text{ mol } H_2O} = 10.4 \text{ kJ}$

53. $835\text{ kJ}\times\frac{1\text{ mol }H_2O}{40.6\text{ kJ}}\times\frac{18.02\text{ g}}{1\text{ mol }H_2O}=371\text{ g }H_2O$

55. $37.4\text{ g }H_2O\times\frac{1\text{ mol }H_2O}{18.02\text{ g}}\times\frac{6.02\text{ kJ}}{1\text{ mol }H_2O}=12.5\text{ kJ}$

57. $34.2\text{ g }H_2O\times\frac{1\text{ mol }H_2O}{18.02\text{ g}}\times\frac{6.02\text{ kJ}}{1\text{ mol }H_2O}=11.4\text{ kJ}$

Intermolecular Forces

59. a) dispersion
    b) dispersion
    c) dispersion, dipole-dipole
    d) dispersion, hydrogen bond, dipole-dipole

61. a) dispersion, dipole-dipole
    b) dispersion, dipole-dipole, hydrogen bond
    c) dispersion
    d) dispersion

63. Choice d would have the highest boiling point because it has a larger molar mass, which indicates stronger dispersion forces.

65. The CH_3OH compound will be a liquid at room temperature because it will have hydrogen bonding, dipole-dipole, and dispersion intermolecular forces, while CH_3SH would contain only dipole-dipole and dispersion intermolecular forces.

67. The first liquid that would start to form is NH_3 because of its ability to form hydrogen bonds. CH_4 cannot form hydrogen bonds, so condensation would not start until a much lower temperature was reached.

69. No, $CH_3CH_2CH_2CH_2CH_3$ is nonpolar where H_2O is polar and they are not miscible.

71. a) No, CCl_4 is nonpolar and H_2O is polar.
    b) Yes, both compounds are nonpolar.
    c) Yes, both compounds are polar.

Types of Solids

73. a) atomic
    b) molecular
    c) ionic
    d) atomic

75. a) molecular
    b) ionic
    c) molecular
    d) molecular

77. Choice c because ionic compounds tend to have high melting points due to the strong electrostatic attraction present in the solid.

79. a) Ti(s) has a higher melting point because metallic atomic solids have stronger metallic bonds compared to the weak dispersion force that holds together non-bonding atomic solids like Ne(s).
    b) $H_2O(s)$ has a higher melting point because of hydrogen bonds, which are not found in H_2S.
    c) Xe(s) has a higher melting point because of stronger dispersion forces found in larger atoms.
    d) NaCl(s) has a higher melting point because electrostatic attraction in ionic solids is stronger than intermolecular forces in covalent solids.

81. $Ne < SO_2 < NH_3 < H_2O < NaF$

Cumulative Problems

83. a) $$78\text{ g H}_2\text{O}\times\frac{1\text{ mol H}_2\text{O}}{18.02\text{ g}}\times\frac{6.02\text{ kJ}}{1\text{ mol H}_2\text{O}}\times\frac{1000\text{ J}}{1\text{ kJ}}=2.6\times10^4\text{J}$$

b) $$78\text{ g H}_2\text{O}\times\frac{1\text{ mol H}_2\text{O}}{18.02\text{ g}}\times\frac{6.02\text{ kJ}}{1\text{ mol H}_2\text{O}}=26\text{ kJ}$$

c) $$78\text{ g H}_2\text{O}\times\frac{1\text{ mol H}_2\text{O}}{18.02\text{ g}}\times\frac{6.02\text{ kJ}}{1\text{ mol H}_2\text{O}}\times\frac{1000\text{ J}}{1\text{ kJ}}\times\frac{1\text{ cal}}{4.18\text{ J}}=6.2\times10^3\text{ cal}$$

d) $$78\text{ g H}_2\text{O}\times\frac{1\text{ mol H}_2\text{O}}{18.02\text{ g}}\times\frac{6.02\text{ kJ}}{1\text{ mol H}_2\text{O}}\times\frac{1000\text{ J}}{1\text{ kJ}}\times\frac{1\text{ cal}}{4.18\text{ J}}\times\frac{1\text{ Cal}}{1000\text{ cal}}=6.2\text{ Cal}$$

85. $$8.5\text{ g H}_2\text{O}\times\frac{1\text{ mol H}_2\text{O}}{18.02\text{ g}}\times\frac{6.02\text{ kJ}}{1\text{ mol H}_2\text{O}}\times\frac{1000\text{ J}}{1\text{ kJ}}=2.8\times10^3\text{ J}$$

$$2.8\times10^3\text{ J}=255\text{g}\times4.18\text{J/g}\cdot{}^\circ\text{C}\times\Delta\text{T}$$

$$\Delta\text{T}=\frac{2.8\times10^3\text{ J}}{255\text{g}\times4.18\text{J/g}\cdot{}^\circ\text{C}}=2.7^\circ\text{C}$$

87. $$q=352\text{g}\times4.18\text{J/g}\cdot{}^\circ\text{C}\times25^\circ\text{C}=3.68\times10^4\text{ J}$$

$$3.68\times10^4\text{ J}\times\frac{1\text{ kJ}}{1000\text{ J}}\times\frac{1\text{ mol H}_2\text{O}}{6.02\text{ kJ}}\times\frac{18.02\text{ g}}{1\text{ mol H}_2\text{O}}=1.1\times10^2\text{g H}_2\text{O}$$

89. Cooling Steam: $18.02\ g\ H_2O \times 1.84\ J/g\cdot{}^\circ C \times 45^\circ C \times \frac{1\ kJ}{1000\ J} = 1.5\ kJ$

Condensing Steam: $1\ mol\ H_2O \times 40.7\ kJ/mol = 40.7\ kJ$

Cooling Water: $18.02\ g\ H_2O \times 4.18\ J/g\bullet{}^\circ C \times 1.00 \times 10^{2\circ}C \times \frac{1\ kJ}{1000\ J} = 7.53\ kJ$

Freezing Water: $1\ mol\ H_2O \times \frac{6.02\ kJ}{1\ mol\ H_2O} = 6.02\ kJ$

Cooling Ice: $18.02\ g\ H_2O \times 2.09\ J/g\bullet{}^\circ C \times 50^\circ C \times \frac{1\ kJ}{1000\ J} = 1.88\ kJ$

$Total = 1.5 + 40.7 + 7.53 + 6.02 + 1.88 = 57.6\ kJ$

91. a) H–S̈e–H Bent Geometry, Dispersion & Dipole-Dipole Forces

b) Ö=S̈–Ö: Bent Geometry, Dispersion & Dipole-Dipole Forces

```
      :C̈l:
       |
c) H – C – C̈l: Tetrahedral Geometry, Dispersion & Dipole-Dipole Forces
       |
      :C̤l:
```

d) Ö=C=Ö Linear Geometry, Dispersion Forces

93. $Na^+[:\ddot{F}:]^-$ $Mg^{2+}[:\ddot{O}:]^{2-}$

Because the strength of a +2 to -2 attraction is stronger than a +1 to -1, MgO has the higher melting point.

95. As the molecular weight increases from Cl to I, the greater the London dispersion forces present which will increase the boiling point as observed. However, HF is the only compound listed that has the ability to form hydrogen bonds, which explains the anomaly in the trend.

97. $\underbrace{\text{Heat lost}}_{\text{Bulk water}} + \underbrace{\text{Heat Gained}}_{\text{Melting ice, Warming water from ice}} = 0$

$$\underbrace{4.18\frac{J}{g\cdot{}^\circ C} \times 550.0\ g\ H_2O \times (T_f - 28.0^\circ C)}_{\text{Heat lost by water}} + \underbrace{23.5\ g\ H_2O \times \frac{1\ mol\ H_2O}{18.02\ g} \times \frac{6.02\times10^3 J}{1\ mol\ H_2O}}_{\text{Heat of fusion of ice (melting the ice)}}$$

$$\underbrace{+\ 4.18\frac{J}{g\cdot{}^\circ C} \times 23.5\ g\ H_2O \times (T_f - 0^\circ C)}_{\text{The water from the newly melted ice cube warms up}} = 0$$

$$(2.30\times10^3 T_f - 6.44\times10^4) + (7.85\times10^3) + (98.2T_f - 0) = 0$$

$$2.40\times10^3 T_f - 5.65\times10^4 = 0 \Rightarrow T_f = \frac{5.65\times10^4}{2.40\times10^3} = 23.5^\circ C$$

Highlight Problems

99. Interior molecules have the most neighbors. The surface molecule is more likely to evaporate. The number of neighbors would change; however, surface molecules will always have less than interior molecules and will always be more likely to evaporate.

101. a) This is a valid criticism because ice displaces more volume than the liquid water that makes it up. The melting of ice in a cup would actually result in a small decrease in volume.
     b) The melting of ice from the continent would increase ocean levels because this water is not currently in the ocean itself.

Solutions

Questions

1. A solution is a homogeneous mixture of two or more substances. Some examples include air, seawater, soda water, and brass.

3. The solvent is the major component of the solution. The solute is the minor component in the solution. An example is soda pop where carbon dioxide and sugar are solutes and water is the solvent. Another example is salt water, in which salt is the solute and water is the solvent.

5. Solubility is the amount of a compound (grams) that will dissolve in a specified amount of a solvent.

7. A strong electrolyte solution is one that will conduct electricity due to the dissociation of ionic species in solution. A nonelectrolyte solution will not conduct electricity due to the absence of ions in solution. Ionic compounds tend to form strong electrolytes and molecular compounds form nonelectrolyte solutions.

9. Recrystallization is the process in which a solid is dissolved in a suitable solvent, usually at elevated temperatures. As the solution cools, it becomes saturated, and the excess solid begins to reform crystals. Recrystallization is a common way to purify a solid. The formation of the crystalline structure tends to reject impurities when re-grown slowly from a saturated solution, resulting in crystals with fewer impurities than the original crystals.

11. The bubbles that form in water when it is heated to a temperature lower than the boiling point are dissolved gases that are coming out of solution. The solubility of gases decreases as a function of increased temperatures.

13. The solubility of a gas increases with increasing pressure and decreases with decreasing pressure. In soda pop, the pressure is provided by a large amount of carbon dioxide gas that is pumped into the can before sealing it. When the can is opened, the pressure is released and the solubility of carbon dioxide decreases, resulting in bubbling.

15. Mass percent is the number of grams of solute per 100 g of solution. Molarity (M) s defined as the number of moles of solute per liter of solution.

17. The addition of a nonvolatile solute will lower the freezing point and raise the boiling point of a solution relative to that of the pure solvent.

19. Molality is a common unit of concentration defined as the moles of solute dissolved per kilogram of solvent.

21. By drinking seawater, the one survivor created a region of high salt concentration on the outside of the cells in his body. This caused osmosis of water out of the cells in the body into the saltwater he had consumed, causing more severe dehydration.

Problems

Solutions

23. a) not a solution
    b) not a solution
    c) solution
    d) solution (Sterling silver is a solid solution of 925 parts Ag and 75 parts Cu.)

25. a) solvent: water, solute: salt
    b) solvent: water, solute: sugar
    c) solvent: water, solute: carbon dioxide

27. a) hexane, ethyl ether, or toluene
    b) water, acetone, or methyl alcohol
    c) hexane, ethyl ether, or toluene
    d) water

Solid Dissolved in Water

29. The dissolved particles are the cations and anions that make up the ionic solute. This solution is referred to as a strong electrolyte.

31. From the graph, the solubility of NaCl at 25 °C is ~35g NaCl/100 g H_2O. A concentration of 35 g NaCl/100 g H_2O is saturated.

33. From the graph, the solubility of KNO_3 at 40°C is ~62 g/100 g H_2O. A concentration of 42 g/100 g H_2O is below the solubility limit, so the solution is initially unsaturated. As the temperature cools from 40 to 0 °C the solubility limit decreases. At a temperature of ~ 28 °C the solubility of KNO_3 drops below the value of 42 g/100 g H_2O. Below this temperature the excess KNO_3 will recrystallize as the solution cools.

35. a) $\frac{30.0\text{ g KClO}_3}{85.0\text{ g H}_2\text{O}} = \frac{35.3\text{ g}}{100\text{ g H}_2\text{O}}$

This is above the saturation limit of $KClO_3$ at 35 °C (~12 g/100 g), so not all the $KClO_3$ will dissolve.

b) $\frac{65.0\text{ g NaNO}_3}{125.0\text{ g H}_2\text{O}} = \frac{52.0\text{ g}}{100\text{ g H}_2\text{O}}$

This is below the saturation limit of $NaNO_3$ at 15 °C (~84 g/100 g), so all the $NaNO_3$ will dissolve.

c) $\frac{32.0\text{ g KCl}}{70.0\text{ g H}_2\text{O}} = \frac{45.7\text{ g}}{100\text{ g H}_2\text{O}}$

This is below the saturation limit of KCl at 82 °C (~52 g/100 g), so all the KCl will dissolve.

Gases Dissolved in Water

37. The solubility of gases decreases as temperature increases. When water is boiled, the dissolved oxygen is completely removed.

39. The solubility of gases (nitrogen) increases with increasing pressure. The diver could prevent this effect either by not diving as deep or by using a helium-oxygen mixture as discussed in Chapter 11.

Mass Percent

41. a) $\text{mass }\% = \frac{41.2\text{ g C}_{12}\text{H}_{22}\text{O}_{11}}{(41.2 + 498)\text{ g solution}} \times 100\% = 7.64\%$

b) $\text{mass }\% = \frac{178\text{ mg C}_6\text{H}_{12}\text{O}_6 \times \frac{1\text{ g}}{1000\text{ mg}}}{(178\text{mg} \times \frac{1\text{ g}}{1000\text{ mg}} + 4.91)\text{ g solution}} \times 100\% = 3.50\%$

c) $\text{mass }\% = \frac{7.55\text{ g NaCl}}{(7.55 + 155)\text{ g solution}} \times 100\% = 4.64\%$

43. $\text{mass }\% = \frac{42\text{ g sugar}}{(42 + 311)\text{ g solution}} \times 100\% = 12\%$

45.

Mass Solute	Mass Solvent	Mass Solution	Mass%
15.5	238.1	253.6	6.11%
22.8	167.2	190.0	12.0%
28.8	183.3	212.1	13.6%
56.9	315.2	372.1	15.3%

47. $\frac{3.5\text{ g NaCl}}{100\text{ g solution}} \times 254\text{ g solution} = 8.9\text{ g NaCl}$

49. a) $\frac{3.7\text{ g sucrose}}{100\text{ g solution}} \times 48\text{ g solution} = 1.8\text{ g sucrose}$

b) $\frac{10.2\text{ mg sucrose}}{100\text{ mg solution}} \times 103\text{ mg solution} = 10.5\text{ mg sucrose}$

c) $\frac{14.3\text{ kg sucrose}}{100\text{ kg solution}} \times 3.2\text{ kg solution} = 0.46\text{ kg sucrose}$

51. a) $1.5\text{ g NaCl} \times \frac{100\text{ g solution}}{0.058\text{ g NaCl}} = 2.6 \times 10^3\text{ g solution}$

b) $1.5\text{ g NaCl} \times \frac{100\text{ g solution}}{1.46\text{ g NaCl}} = 1.0 \times 10^2\text{ g solution}$

c) $1.5\text{ g NaCl} \times \frac{100\text{ g solution}}{8.44\text{ g NaCl}} = 18\text{ g solution}$

53. mass of AgCl solution: $4.8\text{ L} \times \frac{1000\text{ mL}}{1\text{ L}} \times \frac{1.01\text{ g}}{1\text{ mL}} = 4.8 \times 10^3\text{ g}$

$4.8 \times 10^3\text{ g solution} \times \frac{3.4\text{ g Ag}}{100\text{ g solution}} = 1.6 \times 10^2\text{ g Ag}$

55. $45.8\text{ g NaCl} \times \frac{100\text{ g solution}}{3.5\text{ g NaCl}} = 1.3 \times 10^3\text{ g solution}$

57. $115\text{ mg Pb} \times \frac{1\text{ g}}{1000\text{ mg}} \times \frac{100\text{ g solution}}{0.0011\text{ g Pb}} \times \frac{1\text{ ml}}{1.0\text{ g}} = 1.0 \times 10^4\text{ mL solution}$

Molarity

59. a) $M = \frac{0.127\text{ mol sucrose}}{655\text{ mL} \times \frac{1\text{ L}}{1000\text{ mL}}} = \frac{0.127\text{ mol sucrose}}{0.655\text{ L}} = 0.194\text{ M sucrose}$

b) $M = \frac{0.205\text{ mol } KNO_3}{0.875\text{ L}} = 0.234\text{ M } KNO_3$

c) $M = \frac{1.1\text{ mol KCl}}{2.7\text{ L}} = 0.41\text{ M KCl}$

61. a) moles: $22.6 \text{ g } C_{12}H_{22}O_{11} \times \frac{1 \text{ mol } C_{12}H_{22}O_6}{342.34 \text{ g}} = 0.0660 \text{ mol } C_{12}H_{22}O_6$

$$M = \frac{0.0660 \text{ mol } C_{12}H_{22}O_{11}}{0.442 \text{ L}} = 0.149 \text{ M } C_{12}H_{22}O_{11}$$

b) moles: $42.6 \text{ g NaCl} \times \frac{1 \text{ mol NaCl}}{58.44 \text{ g}} = 0.729 \text{ mol NaCl}$

$$M = \frac{0.729 \text{ mol NaCl}}{1.58 \text{ L}} = 0.461 \text{ M NaCl}$$

c) moles: $315 \text{ mg } C_6H_{12}O_6 \times \frac{1 \text{ g}}{1 \times 10^3 \text{ mg}} \times \frac{1 \text{ mol } C_6H_{12}O_6}{180.18 \text{ g}} = 1.75 \times 10^{-3} \text{ mol } C_6H_{12}O_6$

liters: $58.2 \text{ mL} \times \frac{1 \text{ L}}{1000 \text{ mL}} = 0.0582 \text{ L}$

$$M = \frac{1.75 \times 10^{-3} \text{ mol } C_6H_{12}O_6}{0.0582 \text{ L}} = 0.0300 \text{ M } C_6H_{12}O_6$$

63. $6.8 \text{ g NaCl} \times \frac{1 \text{ mol NaCl}}{58.44 \text{ g}} \times \frac{1}{205 \text{ mL}} \times \frac{1000 \text{ mL}}{1 \text{ L}} = 0.57 \text{ M NaCl}$

65. a) $\frac{1.2 \text{ mol NaCl}}{\text{L}} \times 1.5 \text{ L} = 1.8 \text{ mol NaCl}$

b) $\frac{0.85 \text{ mol}}{\text{L}} \times 0.448 \text{ L} = 0.38 \text{ mol NaCl}$

c) $\frac{1.65 \text{ mol NaCl}}{\text{L}} \times 144 \text{ mL} \times \frac{1 \text{ L}}{1000 \text{ mL}} = 0.238 \text{ mol NaCl}$

67. a) $0.15 \text{ mol KCl} \times \frac{1 \text{ L}}{0.255 \text{ mol KCl}} = 0.59 \text{ L}$

b) $0.15 \text{ mol KCl} \times \frac{1 \text{ L}}{1.8 \text{ mol KCl}} = 0.083 \text{ L}$

c) $0.15 \text{ mol KCl} \times \frac{1 \text{ L}}{0.995 \text{ mol KCl}} = 0.15 \text{ L}$

69.

Solute	Mass Solute	Mol Solute	Vol. Soln.	Molarity
KNO_3	22.5 g	0.223 mol	125 mL	1.78 M
$NaHCO_3$	2.10 g	0.0250 mol	250.0 mL	0.100 M
$C_{12}H_{22}O_{11}$	55.38 g	0.162 mol	1.08 L	0.150 M

71. $35\text{ mL} \times \frac{1\text{ L}}{1000\text{ mL}} \times \frac{1.3\text{ mol NaCl}}{1\text{ L}} \times \frac{58.44\text{ g}}{1\text{ mol NaCl}} = 2.7\text{ g NaCl}$

73. $2.5\text{ L} \times \frac{0.100\text{ mol KCl}}{1\text{ L}} \times \frac{74.55\text{ g}}{1\text{ mol KCl}} = 19\text{ g KCl}$

75. $1.5\text{ kg } C_{12}H_{22}O_{11} \times \frac{1000\text{ g}}{1\text{ kg}} \times \frac{1\text{ mol } C_{12}H_{22}O_{11}}{342.34\text{ g}} \times \frac{1\text{ L}}{0.500\text{ mol}} = 8.8\text{ L}$

77. a) 1 mole Cl^- per 1 mole NaCl, $[Cl^-] = 0.15$ M
b) 2 mole Cl^- per 1 mole $CuCl_2$, $[Cl^-] = 0.30$ M
c) 3 mole Cl^- per 1 mole $AlCl_3$, $[Cl^-] = 0.45$ M

79. a) 0.12 M $Na_2SO_4 \Rightarrow$ 0.24 M Na^+, 0.12 M SO_4^{2-}
b) 0.25 M $K_2CO_3 \Rightarrow$ 0.50 M K^+, 0.25 M CO_3^{2-}
c) 0.11 M RbBr $\Rightarrow$ 0.11 M Rb^+, 0.11 M Br^-

Solution Dilution

81. $M_1V_1 = M_2V_2 \Rightarrow (1.2\text{ M})(122\text{ mL}) = (M_2)(500.0\text{ mL})$

$M_2 = \frac{(1.2\text{ M})(122\text{ mL})}{(500.0\text{ mL})} = 0.29\text{ M}$

83. $M_1V_1 = M_2V_2 \Rightarrow (5.5\text{ M})(V_1) = (0.100\text{ M})(2.5\text{ L})$

$V_1 = \frac{(0.100\text{ M})(2.5\text{ L})}{(5.5\text{ M})} = 0.045\text{ L}$

You would dilute 45 mL (0.045 L) of the 5.5 M stock solution to a final volume of 2.5 L.

85. $M_1V_1 = M_2V_2 \Rightarrow (12\text{ M})(25\text{ mL}) = (0.500\text{ M})(V_2)$

$V_2 = \frac{(12\text{ M})(25\text{ mL})}{(0.500\text{ M})} = 6.0 \times 10^2\text{ mL}$

87. $M_1V_1 = M_2V_2 \Rightarrow (12.0\text{ M})(V_1) = (0.250\text{ M})(850.0\text{ mL})$

$V_1 = \frac{(0.250\text{ M})(850.0\text{ mL})}{(12.0\text{ M})} = 17.7\text{ mL}$

89. a) $\frac{0.150 \text{ mol HCl}}{1 \text{ L}} \times 25 \text{ mL} \times \frac{1 \text{ L}}{1000 \text{ mL}} \times \frac{1 \text{ mol NaOH}}{1 \text{ mol HCl}} \times \frac{1 \text{ L}}{0.150 \text{ mol NaOH}}$
$= 0.025 \text{ L}$

b) $\frac{0.055 \text{ mol HCl}}{1 \text{ L}} \times 55 \text{ mL} \times \frac{1 \text{ L}}{1000 \text{ mL}} \times \frac{1 \text{ mol NaOH}}{1 \text{ mol HCl}} \times \frac{1 \text{ L}}{0.150 \text{ mol NaOH}}$
$= 0.020 \text{ L}$

c) $\frac{0.885 \text{ mol HCl}}{1 \text{ L}} \times 175 \text{ mL} \times \frac{1 \text{ L}}{1000 \text{ mL}} \times \frac{1 \text{ mol NaOH}}{1 \text{ mol HCl}} \times \frac{1 \text{ L}}{0.150 \text{ mol NaOH}}$
$= 1.03 \text{ L}$

91. $\frac{0.0112 \text{mol NiCl}_2}{1 \text{ L}} \times 134 \text{ mL} \times \frac{1 \text{ L}}{1000 \text{ mL}} \times \frac{2 \text{ mol K}_3\text{PO}_4}{3 \text{ mol NiCl}_2} \times \frac{1 \text{ L}}{0.225 \text{mol K}_3\text{PO}_4}$
$= 0.00445 \text{ L}$

93. $\frac{0.100 \text{ mol KOH}}{1 \text{ L}} \times 112 \text{ mL} \times \frac{1 \text{ L}}{1000 \text{ mL}} \times \frac{1 \text{ mol H}_3\text{PO}_4}{3 \text{ mol KOH}} \times \frac{1}{10.0 \text{ ml}} \times \frac{1000 \text{ ml}}{1 \text{ L}}$
$= 0.373 \text{ M H}_3\text{PO}_4$

95. $15.0 \text{ g H}_2 \times \frac{1 \text{ mol H}_2}{2.02 \text{ g}} \times \frac{3 \text{ mol H}_2\text{SO}_4}{3 \text{ mol H}_2} \times \frac{1 \text{ L}}{6.0 \text{ mol H}_2\text{SO}_4} = 1.2 \text{ L}$

Molality, Freezing Point Depression, and Boiling Point Elevation

97. a) $\text{m} = \frac{0.25 \text{ mol}}{0.250 \text{ kg}} = 1.0 \text{ m}$

b) $\text{m} = \frac{0.882 \text{ mol}}{0.225 \text{ kg}} = 3.92 \text{ m}$

c) $\text{m} = \frac{0.012 \text{ mol}}{23.1 \text{ g}} \times \frac{1000 \text{ g}}{1 \text{ kg}} = 0.52 \text{ m}$

99. $12.5 \text{ g C}_2\text{H}_6\text{O}_2 \times \frac{1 \text{ mol C}_2\text{H}_6\text{O}_2}{62.08 \text{ g}} \times \frac{1}{135 \text{ g H}_2\text{O}} \times \frac{1000 \text{ g}}{1 \text{ kg}} = 1.49 \text{ m C}_2\text{H}_6\text{O}_2$

101. a) $\Delta T_f = 0.85\ \frac{mol}{kg} \times 1.86 \frac{°C \cdot kg}{mol} = 1.6°C \Rightarrow$ Freezing Point = -1.6°C

b) $\Delta T_f = 1.45\ \frac{mol}{kg} \times 1.86 \frac{°C \cdot kg}{mol} = 2.70°C \Rightarrow$ Freezing Point = -2.70°C

c) $\Delta T_f = 4.8\ \frac{mol}{kg} \times 1.86 \frac{°C \cdot kg}{mol} = 8.9°C \Rightarrow$ Freezing Point = -8.9°C

d) $\Delta T_f = 2.35\ \frac{mol}{kg} \times 1.86 \frac{°C \cdot kg}{mol} = 4.37°C \Rightarrow$ Freezing Point = -4.37°C

103. a) $\Delta T_b = 0.118 \frac{mol}{kg} \times 0.512 \frac{°C \cdot kg}{mol} = 0.0604°C \Rightarrow$ Boiling Point=100.060°C

b) $\Delta T_b = 1.94\ \frac{mol}{kg} \times 0.512 \frac{°C \cdot kg}{mol} = 0.993°C \Rightarrow$ Boiling Point = 100.993°C

c) $\Delta T_b = 3.88\ \frac{mol}{kg} \times 0.512 \frac{°C \cdot kg}{mol} = 1.99°C \Rightarrow$ Boiling Point = 101.99°C

d) $\Delta T_b = 2.16\ \frac{mol}{kg} \times 0.512 \frac{°C \cdot kg}{mol} = 1.11°C \Rightarrow$ Boiling Point = 101.11°C

105. molality: $55.8\text{ g } C_6H_{12}O_6 \times \frac{1\text{ mol } C_6H_{12}O_6}{180.2\text{ g}} \times \frac{1}{455\text{ g}} \times \frac{1000\text{ g}}{1\text{ kg}} = 0.681\text{ m}$

$\Delta T_f = 0.681\ \frac{mol}{kg} \times 1.86 \frac{°C \cdot kg}{mol} = 1.27°C$

Freezing Point= $0.00 - 1.27 = -1.27°C$

$\Delta T_b = 0.681\ \frac{mol}{kg} \times 0.512 \frac{°C \cdot kg}{mol} = 0.349°C$

Boiling Point $= 100.000 + 0.349 = 100.349°C$

Cumulative Problems

107. Molarity: $133\text{ g NaCl} \times \frac{1\text{ mol NaCl}}{58.44\text{ g}} \times \frac{1}{1.00\text{ L}} = 2.28\text{ M}$

Mass Percent: $\frac{133\text{ g}}{1.00\text{ L} \times \frac{1000\text{ ml}}{1\text{ L}} \times \frac{1.08\text{ g}}{1\text{ mL}}} \times 100\% = 12.3\%$

109. $(8.5\ \text{M})(0.125\ \text{L}) = (\text{M}_2)(2.5\ \text{L}) \Rightarrow \text{M}_2 = \dfrac{(8.5\ \text{M})(0.125\ \text{L})}{(2.5\ \text{L})} = 0.43\ \text{M}$

$$10.8\ \text{g NaCl} \times \frac{1\ \text{mol NaCl}}{58.44\ \text{g}} \times \frac{1\ \text{L}}{0.43\ \text{mol NaCl}} = 0.43\ \text{L}$$

111. $3.25\ \text{g KI} \times \dfrac{1\ \text{mol KI}}{166.0\ \text{g}} \times \dfrac{1}{0.0250\ \text{L}} = 0.783\ \text{M KI}$

$$(5.00\ \text{M})(50.00\ \text{mL}) = (0.783\ \text{M})(\text{V}_2) \Rightarrow \text{V}_2 = \frac{(5.00\ \text{M})(50.00\ \text{mL})}{(0.783\ \text{M})} = 319\ \text{mL}$$

113. $\dfrac{5.88\ \text{g NaCl}}{100\ \text{g Soln}} \times \dfrac{1.02\text{g Soln}}{1\ \text{mL}} \times \dfrac{1000\ \text{mL}}{1\ \text{L}} \times \dfrac{1\ \text{mol NaCl}}{58.44\ \text{g}} = 1.03\ \text{M NaCl}$

115. $15.0\ \text{L H}_2 \times \dfrac{1\ \text{mol H}_2}{22.4\ \text{L}} \times \dfrac{3\ \text{mol H}_2\text{SO}_4}{3\ \text{mol H}_2} \times \dfrac{1\ \text{L}}{4.0\ \text{mol}} = 0.17\ \text{L}$

117. $\text{NaCl(aq)} + \text{AgNO}_3\text{(aq)} \rightarrow \text{AgCl(s)} + \text{NaNO}_3\text{(aq)}$

$$\frac{0.45\ \text{mol AgNO}_3}{1\text{L}} \times 0.025\ \text{L} \times \frac{1\ \text{mol NaCl}}{1\ \text{mol AgNO}_3} \times \frac{1\ \text{L}}{1.25\ \text{mol NaCl}} \times \frac{1000\ \text{mL}}{1\ \text{L}} = 9.0\ \text{mL}$$

119. $\dfrac{70.3\ \text{g HNO}_3}{100\ \text{g Soln}} \times \dfrac{1.41\ \text{g Soln}}{1\ \text{mL}} \times \dfrac{1000\ \text{mL}}{1\ \text{L}} \times \dfrac{1\ \text{mol HNO}_3}{63.02\ \text{g}} = 15.7\ \text{M HNO}_3$

$$\text{M}_1\text{V}_1 = \text{M}_2\text{V}_2 \Rightarrow (15.7\ \text{M})(\text{V}_1) = (0.500\ \text{M})(2.5\ \text{L})$$

$$\text{V}_1 = \frac{(0.500\ \text{M})(2.5\ \text{L})}{(15.7\ \text{M})} = 0.080\text{L}$$

$$0.080\ \text{L} \times \frac{1000\ \text{mL}}{1\ \text{L}} = 8.0 \times 10^1\ \text{L}$$

121. moles solute: $58.5\text{ g } C_2H_6O_2 \times \frac{1\text{ mol } C_2H_6O_2}{62.08\text{ g}} = 0.942\text{ mol } C_2H_6O_2$

mass solution: $500.0\text{ ml} \times \frac{1.09\text{ g}}{\text{mL}} = 545\text{ g solution}$

mass solvent: 545 g solution − 58.5 g solute = 486.5 g = 0.487 kg solvent

molality: $\text{m} = \frac{0.942\text{ mol } C_2H_6O_2}{0.487\text{ kg solvent}} = 1.93\text{ m}$

$\Delta T_f = (1.93\text{ m}) \times \left(\frac{1.86\text{ °C}}{\text{m}}\right) = 3.59\text{°C} \Rightarrow \quad T_f = 0.00 - 3.59 = -3.59\text{ °C}$

$\Delta T_b = (1.93\text{ m}) \times \left(\frac{0.512\text{ °C}}{\text{m}}\right) = 0.988\text{°C} \Rightarrow T_b = 100.000 + 0.988 = 100.988\text{ °C}$

123. moles solute: $0.2500\text{ L} \times \frac{5.00\text{ mol}}{\text{L}} = 1.25\text{ mol } C_6H_{12}O_6$

mass solute: $1.25\text{ moles} \times \frac{180.18\text{ g}}{\text{mol}} = 22\underline{5}.23\text{ g}$

mass solution: $1.40\text{ L} \times \frac{1.06\text{ g}}{\text{mL}} \times \frac{1000\text{ mL}}{\text{L}} = 1484\text{ g solution}$

mass solvent: 1484 g solution − 225 g solute = 1259 g = 1.259 kg solvent

molality: $\text{m} = \frac{1.25\text{ mol } C_6H_{12}O_6}{1.259\text{ kg solvent}} = 0.993\text{ m}$

$\Delta T_f = (0.993\text{ m}) \times \left(\frac{1.86\text{ °C}}{\text{m}}\right) = 1.85\text{°C} \Rightarrow T_f = 0.00 - 1.85 = -1.85\text{ °C}$

$\Delta T_b = (0.993\text{ m}) \times \left(\frac{0.512\text{ °C}}{\text{m}}\right) = 0.508\text{°C} \Rightarrow T_b = 100.000 + 0.508 = 100.508\text{ °C}$

125. $\frac{17.5\text{ g/MW}}{0.100\text{ kg}} \times 1.86 \frac{\text{°C} \cdot \text{kg}}{\text{mol}} = 1.8\text{°C} \Rightarrow \frac{17.5\text{ g}}{\text{MW}} = 0.0968\text{mol} \Rightarrow$

$\text{MW} = \frac{17.5\text{ g}}{0.0968\text{ mol}} = 1.80 \times 10^2\text{ g/mol}$

127. $\Delta T_f = m \times 1.86 \frac{\text{°C} \cdot \text{kg}}{\text{mol}} = 6.7\text{°C} \Rightarrow m = \frac{1.86\text{°C} \cdot \text{kg/mol}}{6.7\text{°C}} = 3.60\ m$

$\Delta T_b = 3.60\ m \times 0.512 \frac{\text{°C} \cdot \text{kg}}{\text{mol}} = 1.84\text{°C} \Rightarrow \text{Boiling Point} = 101.84\text{°C}$

129. X= mass glucose, molar mass = 180.18 g/mol

125-X= mass of sucrose, molar mass = 342.34g/mol

$$1.75°\text{C} = \frac{\overbrace{\frac{X}{180.18} + \frac{125\text{-}X}{342.34}}^{\text{Total Moles}}}{0.500\text{ kg}} \times 1.86\frac{°\text{C}\cdot\text{kg}}{\text{mol}} \Rightarrow$$

$$\frac{1.75°\text{C }(0.500\text{ kg})}{1.86\frac{°\text{C}\cdot\text{kg}}{\text{mol}}} = \frac{X}{180.18} + \frac{125}{342.34} - \frac{X}{342.34} \Rightarrow$$

$$\frac{1.75°\text{C }(0.500\text{ kg})}{1.86\frac{°\text{C}\cdot\text{kg}}{\text{mol}}} - \frac{125}{342.34} = \frac{X}{180.18} - \frac{X}{342.34} \Rightarrow$$

$$0.1053 = 0.002629\text{ X}$$

$$\text{Mass of Glucose} = X = 40.1\text{ g}$$

$$\text{Mass of Sucrose} = 125\text{-}X = 84.9\text{g}$$

Highlight Problems

131. a) left to right
b) right to left
c) no movement

133. $0.100\text{ g Hg} \times \frac{1\text{ L}}{0.004\text{ mg}} \times \frac{1000\text{ mg}}{1\text{ g}} = 3\times10^{4}\text{L}$

Acids and Bases

14

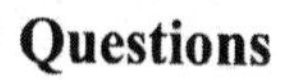

Questions

1. The sour taste is due to the presence of acids in the candy, specifically citric acid and tartaric acid.

3. The main component of stomach acid is hydrochloric acid. The role of acid in our stomach is to kill bacteria and to start breaking food down.

5. The properties of bases are:
   1) Bitter taste
   2) Slippery feel
   3) Turn litmus paper blue
   Some products that contain bases include drain clog removers such as Drano, baking soda and antacids for acid indigestion.

7. An Arrhenius acid produces H+ in aqueous solution. $HCl(aq) \rightarrow H^+(aq) + Cl^-(aq)$

9. A Brønsted-Lowry acid is a proton (H^+) donor. $HCl(aq)+H_2O(l)\rightarrow H_3O^+(aq)+Cl^-(aq)$
   A Brønsted-Lowry base is a proton acceptor. $NH_3(aq)+H_2O(l)\rightarrow NH_4^+(aq)+OH^-(aq)$

11. An acid-base neutralization reaction is when the hydrogen ion from the acid reacts with the hydroxide ion from the base to form water.
    $HCl\ (aq) + NaOH(aq) \rightarrow H_2O(l) + NaOH(aq)$

13. $2HCl(aq) + K_2O(s) \rightarrow H_2O(l) + 2KCl(aq)$

15. A titration is a laboratory procedure in which a reactant of known concentration reacts with another of unknown concentration. The volumes of each solution are carefully monitored until the equivalence point is reached, usually when an indicator causes the reaction mixture to change colors. The equivalence point is experimental point when an exact stoichiometric amount of each reactant has been added.

17. A strong acid will completely dissociate into component ions in solution to form a strong electrolyte. A weak acid will partially dissociate into component ions in solution to form a weak electrolyte.

19. A monoprotic acid contains only one hydrogen ion that will dissociate in solution. A diprotic acid contains two hydrogen ions that will dissociate in solution.

21. Yes, pure water contains H_3O^+ because of the process of self-ionization.

23. a) $[H_3O^+] > 1.0x10^{-7}$ and $[OH^-] < 1.0x10^{-7}$
    b) $[H_3O^+] < 1.0x10^{-7}$ and $[OH^-] > 1.0x10^{-7}$
    c) $[H_3O^+] = 1.0x10^{-7}$ and $[OH^-] = 1.0x10^{-7}$

25. pOH =-log $[OH^-]$; A change of 2.0 pOH units corresponds to 100x change in $[OH^-]$.

27. A buffer is a solution that will resist a change in pH as the weak acid can react with added base and the conjugate base can react with added acid.

29. The cause of acid rain is the formation of SO_2 and NO_2 during the combustion of fossil fuels which will react with atmospheric water and oxygen to produce H_2SO_4 and HNO_3, respectively.

31. Acid rain damages structures made out of metal, marble, cement, and limestone as well as causing harm and possible death to aquatic life and trees.

Problems

Acid and Base Definitions

33. a) acid: $H_2SO_4(aq) \rightarrow H^+(aq) + HSO_4^-(aq)$
    b) base: $Sr(OH)_2(aq) \rightarrow Sr^{2+}(aq) + 2OH^-(aq)$
    c) acid: $HBr(aq) \rightarrow H^+(aq) + Br^-(aq)$
    d) base: $NaOH(aq) \rightarrow Na^+(aq) + OH^-(aq)$

35.

	B-L Acid	B-L Base	Conj. Acid	Conj. Base
a)	HBr	H_2O	H_3O^+	Br^-
b)	H_2O	NH_3	NH_4^+	OH^-
c)	HNO_3	H_2O	H_3O^+	NO_3^-
d)	H_2O	C_5H_5N	$C_5H_5NH^+$	OH^-

37. a) conjugate acid-base pairs
    b) not conjugate acid-base pairs
    c) conjugate acid-base pairs
    d) not conjugate acid-base pairs

39. a) Cl^-
    b) HSO_3^-
    c) CHO_2^-
    d) F^-

41. a) NH_4+
b) $HClO_4$
c) H_2SO_4
d) HCO_3^-

Acid-Base Reactions

43. a) $HI(aq) + NaOH(aq) \rightarrow H_2O(l) + NaI(aq)$
b) $HBr(aq) + KOH(aq) \rightarrow H_2O(l) + KBr(aq)$
c) $2HNO_3(aq) + Ba(OH)_2(aq) \rightarrow 2H_2O(l) + Ba(NO_3)_2(aq)$
d) $2HClO_4(aq) + Sr(OH)_2(aq) \rightarrow 2H_2O(l) + Sr(ClO_4)_2(aq)$

45. a) $2Rb(s) + 2HBr(aq) \rightarrow H_2(g) + 2RbBr(aq)$
b) $Mg(s) + 2HBr(aq) \rightarrow H_2(g) + MgBr_2(aq)$
c) $Ba(s) + 2HBr(aq) \rightarrow H_2(g) + BaBr_2(aq)$
d) $2Al(s) + 6HBr(aq) \rightarrow 3H_2(g) + 2AlCl_3(aq)$

47. a) $MgO(s) + 2HI(aq) \rightarrow H_2O(l) + MgI_2(aq)$
b) $K_2O(s) + 2HI(aq) \rightarrow H_2O(l) + 2KI(aq)$
c) $Rb_2O(s) + 2HI(aq) \rightarrow H_2O(l) + 2RbI(aq)$
d) $CaO(s) + 2HI(aq) \rightarrow H_2O(l) + CaI_2(aq)$

49. a) $6HClO_4(aq) + Fe_2O_3(s) \rightarrow 2Fe(ClO_4)_3(aq) + 3H_2O(l)$
b) $H_2SO_4(aq) + Sr(s) \rightarrow SrSO_4(aq) + H_2(g)$
c) $H_3PO_4(aq) + 3KOH(aq) \rightarrow 3H_2O(l) + K_3PO_4(aq)$

Acid-Base Titrations

51. $HCl(aq) + NaOH(aq) \rightarrow H_2O(l) + NaCl(aq)$

a) $0.02844 \text{ L NaOH} \times \frac{0.1231 \text{ mol NaOH}}{1 \text{ L}} \times \frac{1 \text{ mol HCl}}{1 \text{ mol NaOH}} \times \frac{1}{0.02500 \text{ L HCl}} =$

0.1400 M HCl

b) $0.02122 \text{ L NaOH} \times \frac{0.0972 \text{ mol NaOH}}{1 \text{ L}} \times \frac{1 \text{ mol HCl}}{1 \text{ mol NaOH}} \times \frac{1}{0.01500 \text{ L HCl}} =$

0.138 M HCl

c) $0.01488 \text{ L NaOH} \times \frac{0.1178 \text{ mol NaOH}}{1 \text{ L}} \times \frac{1 \text{ mol HCl}}{1 \text{ mol NaOH}} \times \frac{1}{0.02000 \text{ L HCl}} =$

0.08764 M HCl

d) $0.00688 \text{ L NaOH} \times \frac{0.1325 \text{ mol NaOH}}{1 \text{ L}} \times \frac{1 \text{ mol HCl}}{1 \text{ mol NaOH}} \times \frac{1}{0.00500 \text{ L HCl}} =$

0.182 M HCl

53. $H_2SO_4(aq) + 2KOH(aq) \rightarrow 2H_2O(l) + K_2SO_4(aq)$

$$0.04122\ \text{L KOH} \times \frac{0.1322\ \text{mol KOH}}{1\ \text{L}} \times \frac{1\ \text{mol } H_2SO_4}{2\ \text{mol KOH}} \times \frac{1}{0.02500\ \text{L } H_2SO_4} =$$

$0.1090\ \text{M } H_2SO_4$

55. $H_2SO_4(aq) + 2NaOH(aq) \rightarrow 2H_2O(l) + Na_2SO_4(aq)$

$$0.0100\ \text{L} H_2SO_4 \times \frac{0.102\ \text{mol } H_2SO_4}{1\ \text{L}} \times \frac{2\ \text{mol NaOH}}{1\ \text{mol } H_2SO_4} \times \frac{1\ \text{L}}{0.121\ \text{mol NaOH}} \times \frac{1000\ \text{mL}}{1\ \text{L}}$$

$= 16.9\ \text{mL}$

Strong and Weak Acids and Bases

57. a) strong
 b) weak
 c) strong
 d) weak

59. a) $[H_3O^+] = 1.7$ M
 b) $[H_3O^+] = 1.5$ M
 c) $[H_3O^+] < 0.38$ M
 d) $[H_3O^+] < 1.75$ M

61. a) strong
 b) weak
 c) strong
 d) weak

63. a) $[OH^-] = 0.25$ M
 b) $[OH^-] < 0.25$ M
 c) $[OH^-] = 0.50$ M
 d) $[OH^-] = 1.25$ M

Acidity, Basicity, and K_w

65. a) acidic
 b) acidic
 c) neutral
 d) basic

67. a) $K_w = [H_3O^+][OH^-] \Rightarrow [OH^-] = \frac{K_w}{[H_3O^+]} = \frac{1.0\times10^{-14}}{1.5\times10^{-9}} = 6.7\times10^{-6}\,M$, Basic

b) $K_w = [H_3O^+][OH^-] \Rightarrow [OH^-] = \frac{K_w}{[H_3O^+]} = \frac{1.0\times10^{-14}}{9.3\times10^{-9}} = 1.1\times10^{-6}\,M$, Basic

c) $K_w = [H_3O^+][OH^-] \Rightarrow [OH^-] = \frac{K_w}{[H_3O^+]} = \frac{1.0\times10^{-14}}{2.2\times10^{-6}} = 4.5\times10^{-9}\,M$, Acidic

d) $K_w = [H_3O^+][OH^-] \Rightarrow [OH^-] = \frac{K_w}{[H_3O^+]} = \frac{1.0\times10^{-14}}{7.4\times10^{-4}} = 1.4\times10^{-11}\,M$, Acidic

69. a) $K_w = [H_3O^+][OH^-] \Rightarrow [H_3O^+] = \frac{K_w}{[OH^-]} = \frac{1.0\times10^{-14}}{2.7\times10^{-12}} = 3.7\times10^{-3}\,M$, Acidic

b) $K_w = [H_3O^+][OH^-] \Rightarrow [H_3O^+] = \frac{K_w}{[OH^-]} = \frac{1.0\times10^{-14}}{2.5\times10^{-2}} = 4.0\times10^{-13}\,M$, Basic

c) $K_w = [H_3O^+][OH^-] \Rightarrow [H_3O^+] = \frac{K_w}{[OH^-]} = \frac{1.0\times10^{-14}}{1.1\times10^{-10}} = 9.1\times10^{-5}\,M$, Acidic

d) $K_w = [H_3O^+][OH^-] \Rightarrow [H_3O^+] = \frac{K_w}{[OH^-]} = \frac{1.0\times10^{-14}}{3.3\times10^{-4}} = 3.0\times10^{-11}\,M$, Basic

pH

71. a) basic
b) neutral
c) acidic
d) acidic

73. a) $pH = -\log[H^+] = -\log[1.7\times10^{-8}] = 7.77$
b) $pH = -\log[H^+] = -\log[1.0\times10^{-7}] = 7.00$
c) $pH = -\log[H^+] = -\log[2.2\times10^{-6}] = 5.66$
d) $pH = -\log[H^+] = -\log[7.4\times10^{-4}] = 3.13$

75. a) $[H_3O^+] = 10^{-pH} = 10^{-8.55} = 2.8\times10^{-9}\,M$
b) $[H_3O^+] = 10^{-pH} = 10^{-11.23} = 5.9\times10^{-12}\,M$
c) $[H_3O^+] = 10^{-pH} = 10^{-2.87} = 1.3\times10^{-3}\,M$
d) $[H_3O^+] = 10^{-pH} = 10^{-1.22} = 6.0\times10^{-2}\,M$

77. a) $[H_3O^+] = \frac{K_w}{[OH^-]} = \frac{1.0\times10^{-14}}{1.9\times10^{-7}} = 5.3\times10^{-8}\,M,\ pH = -\log 5.3\times10^{-8} = 7.28$

b) $[H_3O^+] = \frac{K_w}{[OH^-]} = \frac{1.0\times10^{-14}}{2.6\times10^{-8}} = 3.8\times10^{-7}\,M,\ pH = -\log 3.8\times10^{-7} = 6.41$

c) $[H_3O^+] = \frac{K_w}{[OH^-]} = \frac{1.0\times10^{-14}}{7.2\times10^{-11}} = 1.4\times10^{-4}\,M,\ pH = -\log 1.4\times10^{-4} = 3.86$

d) $[H_3O^+] = \frac{K_w}{[OH^-]} = \frac{1.0\times10^{-14}}{9.5\times10^{-2}} = 1.05\times10^{-13}\,M,\ pH = -\log 1.05\times10^{-13} = 12.98$

79. a) $[H_3O^+]=10^{-pH}=10^{-4.25}=5.62\times10^{-5}M,\ [OH^-]=\frac{K_w}{[H_3O^+]}=\frac{1.0\times10^{-14}}{5.62\times10^{-5}}=1.8\times10^{-10}M$

b) $[H_3O^+]=10^{-pH}=10^{-12.53}=2.95\times10^{-13}M,\ [OH^-]=\frac{K_w}{[H_3O^+]}=\frac{1.0\times10^{-14}}{2.95\times10^{-13}}=3.4\times10^{-2}M$

c) $[H_3O^+]=10^{-pH}=10^{-1.50}=3.16\times10^{-2}M,\ [OH^-]=\frac{K_w}{[H_3O^+]}=\frac{1.0\times10^{-14}}{3.16\times10^{-2}}=3.2\times10^{-13}M$

d) $[H_3O^+]=10^{-pH}=10^{-8.25}=5.62\times10^{-9}M,\ [OH^-]=\frac{K_w}{[H_3O^+]}=\frac{1.0\times10^{-14}}{5.62\times10^{-9}}=1.8\times10^{-6}M$

81. a) pH=-log 0.0155 =1.810
b) pOH=-log 1.28×10^{-3}=2.893; pH=14.000-2.893=11.107
c) pH=-log 1.89×10^{-3}=2.724
d) pOH=-log $2(1.54\times10^{-4})$=3.511; pH=14.000-3.511=10.489

pOH

83. a) pOH = -log [1.5×10^{-9}] = 8.82, acidic solution (pH=5.18)
b) pOH = -log [7.0×10^{-5}] = 4.15, basic solution (pH=9.85)
c) pOH = -log [1.0×10^{-7}] = 7.00, neutral solution (pH=7.00)
d) pOH = -log [8.8×10^{-3}] = 2.06, basic solution (pH=11.94)

85. a) pOH = 14 – pH; pOH = 14 – -log [1.2×10^{-8}] = 6.08
b) pOH = 14 – pH; pOH = 14 – -log [5.5×10^{-2}] = 12.74
c) pOH = 14 – pH; pOH = 14 – -log [3.9×10^{-9}] = 5.59
d) pOH = -log [OH^-] =-log [1.88×10^{-13}] = 12.726

87. a) pH = 14 – pOH; pH = 14 – 8.5 = 5.5, acidic
b) pH = 14 – pOH; pH = 14 – 4.2 = 9.8, basic
c) pH = 14 – pOH; pH = 14 – 1.7 = 12.3, basic
d) pH = 14 – OH; pH = 14 – 7.0 = 7.0, neutral

Buffers and Acid Rain

89. Correct answers vary.

91. a) not a buffer
b) not a buffer
c) buffer
d) buffer

93. c) $NaF(aq) + HCl(aq) \rightarrow HF(aq) + NaCl(aq)$
d) $KC_2H_3O_2(aq) + HCl(aq) \rightarrow HC_2H_3O_2(aq) + KCl(aq)$

95. a) $HC_2H_3O_2$
b) NaH_2PO_4
c) $NaCHO_2$

Cumulative Exercises

97. $HCl(aq) + NaOH(aq) \rightarrow H_2O(l) + NaCl(aq)$

$$\frac{0.250 \text{ mol NaOH}}{1 \text{ L}} \times 0.0200 \text{ L} \times \frac{1 \text{ mol HCl}}{1 \text{ mol NaOH}} \times \frac{1 \text{ L}}{0.100 \text{ mol HCl}} = 0.0500 \text{ L}$$

99. $Mg(s) + 2HCl(aq) \rightarrow H_2(g) + MgCl_2(aq)$

$$10.0 \text{ g Mg} \times \frac{1 \text{ mol Mg}}{24.31 \text{ g}} \times \frac{2 \text{ mol HCl}}{1 \text{ mol Mg}} \times \frac{1 \text{ L}}{5.0 \text{ mol HCl}} = 0.16 \text{ L}$$

101. $K_2O(s) + 2HI(aq) \rightarrow H_2O(l) + 2KI(aq)$

$$18.5 \text{ g } K_2O \times \frac{1 \text{ mol } K_2O}{94.20 \text{ g}} \times \frac{2 \text{ mol KI}}{1 \text{ mol } K_2O} \times \frac{166.0 \text{ g}}{1 \text{ mol KI}} = 65.2 \text{ g KI}$$

103. $HX(aq) + NaOH(aq) \rightarrow H_2O(l) + NaX(aq)$

$$\frac{0.1003 \text{ mol NaOH}}{1 \text{ L}} \times 0.02077 \text{ L} \times \frac{1 \text{ mol HX}}{1 \text{ mol NaOH}} = 2.083 \times 10^{-3} \text{ mol}$$

$$\text{molar mass} = \frac{0.125 \text{ g}}{2.083 \times 10^{-3} \text{ mol}} = 60.0 \text{ g/mol}$$

105. $2HCl(aq) + Mg(OH)_2(aq) \rightarrow 2H_2O(l) + MgCl_2(aq)$

$$pH = 1.1 \Rightarrow [H^+] = 10^{-1.1} = 0.079 \text{ M}$$

$$0.400 \text{ g } Mg(OH)_2 \times \frac{1 \text{ mol } Mg(OH)_2}{58.33 \text{g}} \times \frac{2 \text{ mol HCl}}{1 \text{ mol } Mg(OH)_2} \times \frac{1 \text{ L}}{0.079 \text{ mol}} = 0.2 \text{ L}$$

107. a) pH = -log 0.0025 = 2.60; acidic
b) pH = -log 1.8×10^{-12} = 11.74; basic
c) pH = -log 9.6×10^{-9} = 8.02; basic
d) pH = -log 0.0195 = 1.710; acidic

109.

$[H_3O^+]$	$[OH^-]$	pH	pOH	acidic or basic
1.0×10^{-4}	*1.0×10^{-10}*	*4.00*	*10.00*	*acidic*
5.5×10^{-3}	1.8×10^{-12}	2.26	11.74	acidic
3.1×10^{-9}	3.2×10^{-6}	8.51	5.49	basic
4.8×10^{-9}	2.1×10^{-6}	8.32	5.68	basic
2.8×10^{-8}	3.5×10^{-7}	7.55	6.45	basic

111.

	$[H_3O^+]$	$[OH^-]$	pH
a)	[0.0088]	$\frac{1.00 \times 10^{-14}}{0.0088} = 1.1 \times 10^{-12}$	2.06
b)	$[1.5 \times 10^{-3}]$	$\frac{1.00 \times 10^{-14}}{1.5 \times 10^{-3}} = 6.7 \times 10^{-12}$	2.82
c)	$[9.77 \times 10^{-4}]$	$\frac{1.00 \times 10^{-14}}{9.77 \times 10^{-4}} = 1.02 \times 10^{-11}$	3.010
d)	[0.0878]	$\frac{1.00 \times 10^{-14}}{0.0878} = 1.14 \times 10^{-13}$	1.057

113.

	$[H_3O^+]$	$[OH^-]$	pOH	pH
a)	$\frac{1.00 \times 10^{-14}}{0.15} = 6.7 \times 10^{-14}$	[0.15]	0.82	13.18
b)	$\frac{1.00 \times 10^{-14}}{3.0 \times 10^{-3}} = 3.3 \times 10^{-12}$	$[3.0 \times 10^{-3}]$	2.52	11.48
c)	$\frac{1.00 \times 10^{-14}}{9.6 \times 10^{-4}} = 1.0 \times 10^{-11}$	$[9.6 \times 10^{-4}]$	3.02	10.98
d)	$\frac{1.00 \times 10^{-14}}{8.7 \times 10^{-5}} = 1.1 \times 10^{-10}$	$[8.7 \times 10^{-5}]$	4.06	9.94

115. $2HCl(aq) + Fe(s) \rightarrow H_2(g) + FeCl_2(aq)$

$$500.0 \text{ g Fe} \times \frac{1 \text{ mol Fe}}{55.85 \text{ g}} \times \frac{2 \text{ mol HCl}}{1 \text{ mol Fe}} \times \frac{1\text{L}}{12.0 \text{ mol}} = 1.49\text{L}$$

No, a pen cannot contain this amount of acid.

117. $HCl(aq) + NaOH(aq) \rightarrow NaCl(aq) + H_2O(l)$

$$0.125\text{L} \times \frac{0.0250\text{mol HCl}}{1\text{L}} = 0.00313 \text{ mol HCl}$$

$$0.0750\text{L} \times \frac{0.0500\text{mol NaOH}}{1\text{L}} = 0.00375 \text{ mol NaOH}$$

$$0.00375 - 0.00313 = 6.25 \times 10^{-4} \text{mol NaOH excess}$$

$$[OH^-] = \frac{6.25 \times 10^{-4} \text{mol NaOH}}{(0.125 \text{ L} + 0.075 \text{ L})} = 3.13 \times 10^{-3}$$

$$pOH = -\log 3.13 \times 10^{-3} = 2.51; \quad pH = 14 - 2.51 = 11.49$$

119. $$0.050 \text{ mL} \times \frac{1 \text{ L}}{1000 \text{ mL}} \times \frac{1.00 \times 10^{-7} \text{mol H}^+}{1 \text{ L}} \times \frac{6.022 \times 10^{23} \text{ H}^+}{1 \text{ mol H}^+} = 3.0 \times 10^{12} \text{ H}^+ \text{ ions}$$

121. Total $[OH^-] = 10^{-1.51} = 0.031$ M
moles $OH^- = 0.031$ mol $OH^-/L \times 4.00$ L $= 0.124$ mol OH^-
*moles $OH^- = 1 \times$ moles NaOH $+ 2 \times$ moles $Sr(OH)_2 = 0.124$ mol OH^-
mol NaOH + mol $Sr(OH)_2 = 0.100$ mol $\Rightarrow$ *mol NaOH $= 0.100 -$ mol $Sr(OH)_2$
Using the two * equations:
$1 \times (0.100 -$ mol $Sr(OH)_2) + 2 \times$ moles $Sr(OH)_2 = 0.124$ mol OH^-
mol $Sr(OH)_2 = (0.124 - 0.100)$ mol $Sr(OH)_2 = 0.024$mol $Sr(OH)_2$
and 0.076 mol NaOH.

Highlight Problems

123. a) weak
b) strong
c) weak
d) strong

125. Great Lakes: $pH = 4.5 \Rightarrow [H^+] = 10^{-4.5} = 3.2 \times 10^{-5}$M

West Coast: $pH = 5.4 \Rightarrow [H^+] = 10^{-5.4} = 4.0 \times 10^{-6}$M

Ratio: $\frac{3.2 \times 10^{-5}}{4.0 \times 10^{-6}} = 8$ times more concentrated

Chemical Equilibrium

15

Questions

1. The two general concepts involved in equilibrium are sameness and changelessness. A reaction with a fast rate proceeds quickly; a large amount of reactant is converted to product in a certain period of time. A reaction with a slow rate proceeds slowly; only a small amount of reactant is converted to product in the same period of time.

3. Chemists seek to control reaction rates so that a desired result is obtained. The goal may be to prevent a reaction from going too fast and becoming dangerous, or it may be to speed it up so that it proceeds at a fast enough rate to be used in industrial settings.

5. The two factors that influence reaction rates are concentration and temperature. The rate of a reaction increases with increasing concentration. The rate of a reaction increases with increasing temperature.

7. Dynamic equilibrium exists when the rate of the forward reaction is equal to the rate of the reverse reaction.

9. Because the rates of the forward and reverse reactions are the same at equilibrium, the relative concentrations of reactants and products become constant. It doesn't matter if it is a high or a low concentration, it simply remains constant.

11. The equilibrium constant is a measure of how far a reaction goes toward completion. It is significant because it is used to quantify the concentration of all compounds in a reaction at equilibrium.

13. A small equilibrium constant shows that a reverse reaction is favored and, when equilibrium is reached, there will be more reactants than products. A large equilibrium constant shows that a forward reaction is favored and, when equilibrium is reached, there will be more products than reactants.

15. No, the concentrations of reactants and products will not always be the same in every equilibrium mixture of a particular reaction at a given temperature. The final concentrations will depend on the initial concentrations and will adjust accordingly so that the K_{eq} value is achieved.

17. Correct answers may vary.

19. The reaction will proceed in the reverse (left) direction.

21. The reaction will proceed in the forward (right) direction.

23. The reaction will proceed in the forward (right) direction.

25. The reaction will proceed in the reverse (left) direction.

27. The reaction will proceed in the reverse (left) direction if you increase the temperature of an exothermic reaction at equilibrium. The reaction will proceed in the forward (right) direction if you decrease the temperature of an exothermic reaction at equilibrium.

29. $K_{sp}=[A^{2+}][B^{-}]^2$

31. Solubility is the amount of a compound that dissolves in a specified amount of liquid. The molar solubility is just the solubility expressed in molarity.

33. Two reactants with a large K_{eq} may not react immediately when combined because there may be a large activation energy barrier that the reactants must overcome in order for the reaction to occur.

35. The catalyst does NOT affect the value of the equilibrium constant: it merely allows the reaction to reach equilibrium faster than without a catalyst.

Problems

The Rate of Reaction

37. The rate would decrease because the effective concentration of the reactants has been decreased which lowers the rate of a reaction.

39. Reaction rates tend to decrease with decreasing temperature so all life processes (chemical reactions) would have decreased rates.

41. The reaction rate at the second reading would be slower than the initial rate because the reactants are being consumed, thus lowering their concentrations and the resulting reaction rate.

The Equilibrium Constant

43. a) $K_{eq} = \frac{[N_2O_4]}{[NO_2]^2}$

b) $K_{eq} = \frac{[NO]^2[Br_2]}{[BrNO]^2}$

c) $K_{eq} = \frac{[H_2][CO_2]}{[H_2O][CO]}$

d) $K_{eq} = \frac{[CS_2][H_2]^4}{[CH_4][H_2S]^2}$

45. a) $K_{eq} = \frac{[Cl_2]}{[PCl_5]}$

b) $K_{eq} = [O_2]^3$

c) $K_{eq} = \frac{[H_3O^+][F^-]}{[HF]}$

d) $K_{eq} = \frac{[NH_4^+][OH^-]}{[NH_3]}$

47. $K_{eq} = \frac{[H_2]^2[S_2]}{[H_2S]^2}$

49. a) $K_{eq} \gg 1$, products dominate equilibrium
b) $K_{eq} \approx 1$, significant amounts of reactants and products
c) $K_{eq} \ll 1$, reactants dominate equilibrium
d) $K_{eq} \approx 1$, significant amounts of reactants and products

Calculating and Using Equilibrium Constants

51. $K_{eq} = \frac{[0.105][0.0844]}{[0.225]} = 0.0394$

53. $K_{eq} = \frac{[2.74\times10^{-2}]^2[7.54\times10^{-3}]}{[0.562]^2} = 1.79\times10^{-5}$

55. $K_{eq} = [0.278][0.355] = 0.0987$

57. $K_{eq} = \frac{[0.0255][0.135]}{[SbCl_5]} = 4.9\times10^{-4} \Rightarrow [SbCl_5] = \frac{[0.0255][0.135]}{4.9\times10^{-4}} = 7.0\text{ M}$

59. $K_{eq} = \frac{[ICl]^2}{[0.0112][0.0155]} = 81.9 \Rightarrow [ICl] = \sqrt{(81.9)[0.0112][0.0155]} = 0.119$

61.

T(K)	$[N_2]$	$[H_2]$	$[NH_3]$	K_{eq}
500	0.115	0.105	0.439	$\underline{1.45 \times 10^3}$
575	0.110	$\underline{0.25}$	0.128	9.6
775	0.120	0.140	$\underline{4.39 \times 10^{-3}}$	0.0584

Le Châtelier's Principle

63. a) shift right
    b) shift left
    c) shift right

65. a) unchanged
    b) shift left
    c) shift left
    d) shift right

67. a) shift right
    b) shift left

69. a) no effect
    b) no effect

71. a) shift right
    b) shift left

73. a) shift left
    b) shift right

75. a) no effect
    b) shift right
    c) shift left
    d) shift right
    e) no effect

The Solubility-Product Constant

77. a) $CaSO_4 \rightleftarrows Ca^{2+}(aq) + SO_4^{2-}(aq)$; $K_{sp} = \left[Ca^{2+}\right]\left[SO_4^{2-}\right]$

b) $AgCl \rightleftarrows Ag^{+}(aq) + Cl^{-}(aq)$; $K_{sp} = \left[Ag^{+}\right]\left[Cl^{-}\right]$

c) $CuS \rightleftarrows Cu^{2+}(aq) + S^{2-}(aq)$; $K_{sp} = \left[Cu^{2+}\right]\left[S^{2-}\right]$

d) $FeCO_3 \rightleftarrows Fe^{2+}(aq) + CO_3^{2-}(aq)$; $K_{sp} = \left[Fe^{2+}\right]\left[CO_3^{2-}\right]$

79. $K_{sp} = [Fe^{2+}][OH^{-}]^2$

81. $K_{sp} = [2.6\times10^{-4}][5.2\times10^{-4}]^2 = 7.0\times10^{-11}$

83. $K_{sp} = [Pb^{2+}][SO_4^{2-}] \Rightarrow [SO_4^{2-}] = \dfrac{K_{sp}}{[Pb^{2+}]} = \dfrac{1.82\times10^{-8}}{1.35\times10^{-4}} = 1.35\times10^{-4}$

85. $[Ca^{2+}] = [CO_3^{2-}] = S \Rightarrow K_{sp} = S^2 \Rightarrow S = \sqrt{4.96\times10^{-9}} = 7.04\times10^{-5}$ M

87. $[Mg^{2+}] = [CO_3^{2-}] = S \Rightarrow K_{sp} = S^2 \Rightarrow S = \sqrt{6.82\times10^{-6}} = 2.61\times10^{-3}$ M

89.

Compound	[Cation]	[Anion]	K_{sp}
$SrCO_3$	2.4×10^{-5}	2.4×10^{-5}	$\underline{5.8\times10^{-10}}$
SrF_2	1.0×10^{-3}	$\underline{2.0\times10^{-3}}$	4.0×10^{-9}
Ag_2CO_3	$\underline{2.6\times10^{-4}}$	1.3×10^{-4}	8.8×10^{-12}

Cumulative Problems

91. $Fe^{3+}{}_{equilibrium} = Fe^{3+}{}_{initial} - Fe^{3+}{}_{reacted}$

$$\frac{1.7\times10^{-4}\ \text{mol FeSCN}^{2+}}{\text{L}} \times \frac{1\ \text{mol Fe}^{3+}}{1\ \text{mol FeSCN}^{2+}} = 1.7\times10^{-4}\ \text{M Fe}^{2+}$$

$[Fe^{3+}] = 1.0\times10^{-3} - 1.7\times10^{-4} = 8.3\times10^{-4}$

$SCN^{-}{}_{equilibrium} = SCN^{-}{}_{initial} - SCN^{-}{}_{reacted}$

$$\frac{1.7\times10^{-4}\ \text{mol FeSCN}^{2+}}{\text{L}} \times \frac{1\ \text{mol SCN}^{-}}{1\ \text{mol FeSCN}^{2+}} = 1.7\times10^{-4}\ \text{M SCN}^{-}$$

$[SCN^{-}] = 8.0\times10^{-4} - 1.7\times10^{-4} = 6.3\times10^{-4}$

$$K_{eq} = \frac{[FeSCN^{2+}]}{[Fe^{3+}][SCN^{-}]} = \frac{[1.7\times10^{-4}]}{[8.3\times10^{-4}][6.3\times10^{-4}]} = 3.3\times10^{2}$$

93. $K_{eq} = \frac{[HI]^2}{[H_2][I_2]} \Rightarrow 6.17\times10^{-2} = \frac{[HI]^2}{[0.104][0.0202]} \Rightarrow$

$[HI] = \sqrt{(6.17\times10^{-2})[0.104][0.0202]} = 0.0114\ M$

$\frac{0.0114\ mol\ HI}{1\ L} \times 3.67\ L \times \frac{127.91\ g}{1\ mol\ HI} = 5.34\ g\ HI$

95. a) no
b) yes
c) yes
d) yes

97. $[Cu^{2+}]=[S^{2-}]=S \Rightarrow K_{sp} = S^2 \Rightarrow S = \sqrt{1.27\times10^{-36}} = 1.13\times10^{-18}\ M$

$\frac{1.13\times10^{-18}\ mol\ CuS}{1\ L} \times 15.0\ L \times \frac{95.62\ g}{1\ mol\ CuS} = 1.6\times10^{-15} g$

99. $\frac{0.105\ g\ Na_2SO_4}{0.100\ L} \times \frac{1\ mol\ Na_2SO_4}{142.0\ g} = 7.39\times10^{-3} M$

$Q=(0.025)(7.39\times10^{-3}) = 1.85\times10^{-4};\ K_{sp}=7.10\times10^{-5}$

$Q>K_{sp}$ ∴ Yes, a precipitate will form.

101. $\frac{4.15\ g\ CaCrO_4}{1\ L} \times \frac{1\ mol\ CaCrO_4}{156.1\ g} = 0.0266\ M$

$Ksp=[Ca^{2+}][CrO_4^{\ 2-}] = (0.0266)^2 = 7.07\times10^{-4}$

103. $K= [CO_2] = 4.1\times10^{-4} mol/L$

$mol\ CO_2 = 4.1\times10^{-4} mol/L \times 0.500 L = 0.00021\ mol\ CO_2 \Rightarrow$

$0.00021\ mol\ CO_2 \times \frac{1 mol\ CaO}{1 mol\ CO_2} \times \frac{56.08g}{1\ mol\ CaO} = 0.012g\ CaO$

105. $MgCO_3(s) \rightleftarrows Mg^{2+}(aq) + CO_3^{2-}(aq)\ \ K_{sp} = [Mg^{2+}][CO_3^{2-}] = 6.82\times10^{-6}$

$[0.115][CO_3^{2-}] = 6.82\times10^{-6} \Rightarrow [CO_3^{2-}] = \frac{6.82\times10^{-6}}{[0.115]} = 5.93\times10^{-5} M$

$\frac{5.93\times10^{-5}\ mol\ K_2CO_3}{L} \times 2.55\ L \times \frac{138.21\ g}{1\ mol\ K_2CO_3} = 0.0209\ g\ K_2CO_3$

107. Equilibrium was reached at figure e.

109. $[Ca^{2+}]=[CO_3^{2-}] = S \Rightarrow K_{sp} = S^2 \Rightarrow S = \sqrt{4.96\times10^{-9}} = 7.04\times10^{-5}$ M

$$0.250 \text{ g CaCO}_3 \times \frac{1 \text{ mol CaCO}_3}{100.09 \text{ g}} \times \frac{1 \text{ L}}{7.04\times10^{-5} \text{ mol}} = 35.5 \text{ L}$$

Oxidation and Reduction 16

Questions

1. The fuel cells use the electron-gaining tendency of oxygen and the electron-losing tendency of hydrogen to force electrons to move through a wire, creating the electricity that powers the car.

3. a) Oxidation is when a substance gains oxygen atoms in the reaction. Reduction is when oxygen is lost from a substance.
   b) Oxidation is when a substance loses electrons. Reduction is when a substance gains electrons.
   c) Oxidation is an increase in the oxidation state of a substance. Reduction is a decrease in the oxidation state of a substance.

5. Good oxidizing agents have a strong tendency to gain electrons in reactions.

7. The oxidation state of a free element in its naturally occurring state is zero. The oxidation state of a monatomic ion is equal to the charge of the ion.

9. For an ion, the sum of the oxidation states of the individual atoms must sum to the ion charge.

11. In a redox reaction, an atom that undergoes an increase in oxidation state is oxidized. An atom that undergoes a decrease in oxidation state is reduced.

13. When balancing redox equations, the number of electrons lost in the oxidation half-reaction must equal the number of electrons gained in the reduction half-reaction.

15. When balancing aqueous redox reactions, charge is balanced using electrons (e^-).

17. The metals at the top of the activity series are the most reactive.

19. The metals at the bottom of the activity series are least likely to lose electrons.

21. Metals above H_2 on the activity series will dissolve in acids, while metals below H_2 will not dissolve in acids.

23. Oxidation occurs at the anode of an electrochemical cell.

25. The role of a salt bridge is to allow for the flow of ions between two halves of an electrochemical cell, which completes the electrical circuit.

27. The common dry cell battery contains only a small amount of liquid water within it, hence the name. The cell voltage is approximately 1.5 volts.
The anode reaction: $Zn(s) \rightarrow Zn^{2+}(aq) + 2e^-$.
The cathode reaction: $2MnO_2(s) + 2NH_4^+(aq) + 2e^- \rightarrow Mn_2O_3(s) + 2NH_3(g) + H_2O(l)$.

29. A fuel cell operates much like a battery, however, the reactants in a fuel cell are continually replenished from an external supply as the reaction proceeds. The most common fuel cell is the hydrogen-oxygen fuel cell. The reaction at the anode is the oxidation of hydrogen gas to form water according to the half-reaction:
$H_2(g) + 4OH^-(aq) \rightarrow 4H_2O(l) + 4e^-$.
The reaction at the cathode corresponds to the reduction of oxygen gas to form hydroxide ion according to the half-reaction: $O_2(g) + 2H_2O(l) + 4e^- \rightarrow 4OH^-$.

31. The oxidation of metals to form a metal oxide is called corrosion. For the corrosion of iron: The oxidation half-reaction is $2Fe(s) \rightarrow 2Fe^{2+}(aq) + 4e^-$. The reduction half-reaction is $O_2(g) + 2H_2O(l) + 4e^- \rightarrow 4OH^-(aq)$.

Problems

Oxidation and Reduction

33. a) H_2
b) Al
c) Al

35.

	Oxidized	Reduced
a)	Sr	O_2
b)	Ca	Cl_2
c)	Mg	Ni^{2+}

37.

	Oxidizing Agent	Reducing Agent
a)	O_2	Sr
b)	Cl_2	Ca
c)	Ni^{2+}	Mg

39. A good oxidizing agent is a substance that is easily reduced (gains electrons). Those elements that form anions will serve as good oxidizing agents.
a) No
b) Yes
c) No
d) Yes

41. A good reducing agent is a substance that is easily oxidized (loses electrons). Those elements that form cations will serve as good reducing agents.
   a) Yes
   b) No
   c) Yes
   d) No

43.

	Oxidized	Reduced	Oxidizing Agent	Reducing Agent
a)	N_2	O_2	O_2	N_2
b)	C in CO	O_2	O_2	C in CO
c)	Sb in $SbCl_3$	Cl in Cl_2	Cl_2	Sb in $SbCl_3$
d)	K	Pb^{2+}	Pb^{2+}	K

Oxidation States

45. a) 0
   b) +2
   c) +3
   d) 0

47. a) Na=+1, Cl=−1
   b) Ca=+2, F=−1
   c) S=+4, O=−2
   d) H=+1, S=−2

49. a) +2
   b) +4
   c) +1

51. a) C=+4, O=−2
   b) O=−2, H=+1
   c) N=+5, O=−2
   d) N=+3, O=−2

53. a) +1
   b) +3
   c) +5
   d) +7

55. a) Cu=+2, N=+5, O=−2
   b) Sr=+2, O=−2, H=+1
   c) K=+1, O=−2, Cr=+6
   d) Na=+1, H=+1, O=−2, C=+4

57. a) Sb: +5 → +3, reduced; Cl: −1 → 0, oxidized
b) C: +2 → +4, oxidized; Cl: 0 → −1, reduced; O: −2 → −2, neither
c) N: +2 → +3, oxidized; O: −2 → −2, neither; Br: 0 → −1, reduced
d) H: 0 → +1, oxidized; C: +4 → +2, reduced; O: −2 → −2, neither

59. Na: 0 → +1, reducing agent; H: +1 → 0, oxidizing agent; O: −2 → −2, neither

Balancing Redox Reactions

Refer to the following guide when examining the solutions for questions 63-72 to determine what is being done in each line.

1. *Assign oxidation states.*
2. *Separate into half reactions.*
3. *Balance half reactions;*
   - *balance all atoms other than O and H.*
   - *balance O by adding H_2O to side lacking O.*
   - *balance H by adding H^+ to side lacking H.*
4. *Balance charge by adding e- to one side.*
5. *Make # of electrons gained/lost equal by multiplying half reactions by appropriate coefficients.*
6. *Add half reactions & cancel species that appear on both sides.*
7. *(In basic solution) add the number of OH^- equal to the number of H^+ to both sides.*
   - *$OH^- + H^+ \rightarrow H_2O$*
   - *cancel water molecules if on both sides*

61. a) 1: $\underset{0}{K}(s) + \underset{+3}{Cr^{3+}}(aq) \rightarrow \underset{0}{Cr}(s) + \underset{+1}{K^+}(aq)$

2: $K(s) \rightarrow K^+(aq)$ $\qquad$ $Cr^{+3}(aq) \rightarrow Cr(s)$

3: $K(s) \rightarrow K^+(aq)$ $\qquad$ $Cr^{+3}(aq) \rightarrow Cr(s)$

4: $K(s) \rightarrow K^+(aq) + 1e^-$ $\qquad$ $Cr^{+3}(aq) + 3e^- \rightarrow Cr(s)$

5: $3K(s) \rightarrow 3K^+(aq) + 3e^-$ $\qquad$ $Cr^{+3}(aq) + 3e^- \rightarrow Cr(s)$

6: Overall: $3K(s) + Cr^{+3}(aq) \rightarrow 3K^+(aq) + Cr(s)$

b) 1: $\underset{0}{Mg}(s) + \underset{+3}{Cr}^{+3}(aq) \rightarrow \underset{+2}{Mg}^{2+}(aq) + \underset{0}{Cr}(s)$

2: $Mg(s) \rightarrow Mg^{2+}(aq)$ $\quad Cr^{3+}(aq) \rightarrow Cr(s)$

3: $Mg(s) \rightarrow Mg^{2+}(aq)$ $\quad Cr^{3+}(aq) \rightarrow Cr(s)$

4: $Mg(s) \rightarrow Mg^{2+}(aq) + 2e^-$ $\quad Cr^{3+}(aq) + 3e^- \rightarrow Cr(s)$

5: $3Mg(s) \rightarrow 3Mg^{2+}(aq) + 6e^-$ $\quad 2Cr^{3+}(aq) + 6e^- \rightarrow 2Cr(s)$

6: Overall: $3Mg(s) + 2Cr^{3+}(aq) \rightarrow 3Mg^{+2}(aq) + 2Cr(s)$

c) 1: $\underset{0}{Al}(s) + \underset{+2}{Fe}^{2+}(aq) \rightarrow \underset{+3}{Al}^{3+}(aq) + \underset{0}{Fe}(s)$

2: $Al(s) \rightarrow Al^{3+}(aq)$ $\quad Fe^{2+}(aq) \rightarrow Fe(s)$

3: $Al(s) \rightarrow Al^{3+}(aq)$ $\quad Fe^{2+}(aq) \rightarrow Fe(s)$

4: $Al(s) \rightarrow Al^{3+}(aq) + 3e^-$ $\quad Fe^{2+}(aq) + 2e^- \rightarrow Fe(s)$

5: $2Al(s) \rightarrow 2Al^{3+}(aq) + 6e^-$ $\quad 3Fe^{2+}(aq) + 6e^- \rightarrow 3Fe(s)$

6: Overall: $Al(s) + 3Fe^{2+}(aq) \rightarrow Al^{3+}(aq) + 3Fe(s)$

63. a) 1: $\underset{+7}{Mn}O_4^-(aq) \rightarrow \underset{+2}{Mn}^{2+}(aq)$ reduction

2: $MnO_4^-(aq) \rightarrow Mn^{2+}(aq)$

3: $MnO_4^-(aq) + 8H^+(aq) \rightarrow Mn^{2+}(aq) + 4H_2O(l)$

4: $MnO_4^-(aq) + 8H^+(aq) + 5e^- \rightarrow Mn^{2+}(aq) + 4H_2O(l)$

b) 1: $\underset{+2}{Pb}^{2+}(aq) \rightarrow \underset{+4}{Pb}O_2(s)$ oxidation

2: $Pb^{2+}(aq) \rightarrow PbO_2(s)$

3: $Pb^{2+}(aq) + 2H_2O(l) \rightarrow PbO_2(s) + 4H^+(aq)$

4: $Pb^{+2}(aq) + 2H_2O(l) \rightarrow PbO_2(s) + 4H^+(aq) + 2e^-$

c) 1: $\underset{+5}{I}O_3^-(aq) \rightarrow \underset{0}{I_2}(s)$ reduction

2: $IO_3^-(aq) \rightarrow I_2(s)$

3: $2IO_3^-(aq) + 12H^+(aq) \rightarrow I_2(s) + 6H_2O(l)$

4: $2IO_3^-(aq) + 12H^+(aq) + 10e^- \rightarrow I_2(s) + 6H_2O(l)$

d) 1: $\underset{+4}{S}O_2(g) \rightarrow \underset{+6}{S}O_4^{2-}(aq)$ oxidation

2: $SO_2(g) \rightarrow SO_4^{2-}(aq)$

3: $SO_2(g) + 2H_2O(l) \rightarrow SO_4^{2-}(aq) + 4H^+(aq)$

4: $SO_2(g) + 2H_2O(l) \rightarrow SO_4^{2-}(aq) + 4H^+(aq) + 2e^-$

65. a) 1: $\underset{+4}{\text{Pb}}O_2(s) + \underset{-1}{I^-}(aq) \rightarrow \underset{+2}{\text{Pb}^{2+}}(aq) + \underset{0}{I_2}(s)$

2: $PbO_2 \rightarrow Pb^{2+}$ $\quad$ $I^- \rightarrow I_2$

3: $PbO_2 + 4H^+ \rightarrow Pb^{2+} + 2H_2O$ $\quad$ $2I^- \rightarrow I_2$

4: $PbO_2 + 4H^+ + 2e^- \rightarrow Pb^{2+} + 2H_2O$ $\quad$ $2I^- \rightarrow I_2 + 2e^-$

5: $PbO_2 + 4H^+ + 2e^- \rightarrow Pb^{2+} + 2H_2O$ $\quad$ $2I^- \rightarrow I_2 + 2e^-$

6: $PbO_2(s) + 4H^+(aq) + 2I^-(aq) \rightarrow Pb^{2+}(aq) + 2H_2O(l) + I_2(s)$

b) 1: $\underset{+4}{S}O_3^{2-}(aq) + \underset{+7}{\text{Mn}}O_4^-(aq) \rightarrow \underset{+6}{S}O_4^{2-}(aq) + \underset{+2}{\text{Mn}^{2+}}(s)$

2: $SO_3^{2-} \rightarrow SO_4^{2-}$ $\quad$ $MnO_4^- \rightarrow Mn^{2+}$

3: $SO_3^{2-} + H_2O \rightarrow SO_4^{2-} + 2H^+$ $\quad$ $MnO_4^- + 8H^+ \rightarrow Mn^{2+} + 4H_2O$

4: $SO_3^{2-} + H_2O \rightarrow SO_4^{2-} + 2H^+ + 2e^-$ $\quad$ $MnO_4^- + 8H^+ + 5e^- \rightarrow Mn^{2+} + 4H_2O$

5: $5SO_3^{2-} + 5H_2O \rightarrow 5SO_4^{2-} + 10H^+ + 10e^-$ $\quad$ $2MnO_4^- + 16H^+ + 10e^- \rightarrow 2Mn^{2+} + 8H_2O$

6: $5SO_3^{2-}(aq) + 2MnO_4^-(aq) + 6H^+(aq) \rightarrow 5SO_4^{2-}(aq) + 2Mn^{2+}(aq) + 3H_2O(l)$

c) 1: $\underset{+2}{S_2}O_3^{2-}(aq) + \underset{0}{Cl_2}(g) \rightarrow \underset{+6}{S}O_4^{2-}(aq) + \underset{-1}{Cl^-}(aq)$

2: $S_2O_3^{2-} \rightarrow SO_4^{2-}$ $\quad$ $Cl_2 \rightarrow Cl^-$

3: $S_2O_3^{2-} + 5H_2O \rightarrow 2SO_4^{2-} + 10H^+$ $\quad$ $Cl_2 \rightarrow 2Cl^-$

4: $S_2O_3^{2-} + 5H_2O \rightarrow 2SO_4^{2-} + 10H^+ + 8e^-$ $\quad$ $Cl_2 + 2e^- \rightarrow 2Cl^-$

5: $S_2O_3^{2-} + 5H_2O \rightarrow 2SO_4^{2-} + 10H^+ + 8e^-$ $\quad$ $4Cl_2 + 8e^- \rightarrow 8Cl^-$

6: $S_2O_3^{2-}(aq) + 5H_2O(l) + 4Cl_2(g) \rightarrow 2SO_4^{2-}(aq) + 10H^+(aq) + 8Cl^-(aq)$

67. a) 1: $\underset{+7}{Cl}O_4^-(aq) + \underset{-1}{Cl^-}(aq) \rightarrow \underset{+5}{Cl}O_3^-(aq) + \underset{0}{Cl_2}(g)$

2: $ClO_4^- \rightarrow ClO_3^-$ $\quad$ $Cl^- \rightarrow Cl_2$

3: $ClO_4^- + 2H^+ \rightarrow ClO_3^- + H_2O$ $\quad$ $2Cl^- \rightarrow Cl_2$

4: $ClO_4^- + 2H^+ + 2e^- \rightarrow ClO_3^- + H_2O$ $\quad$ $2Cl^- \rightarrow Cl_2 + 2e^-$

5: $ClO_4^- + 2H^+ + 2e^- \rightarrow ClO_3^- + H_2O$ $\quad$ $2Cl^- \rightarrow Cl_2 + 2e^-$

6: $ClO_4^-(aq) + 2H^+(aq) + 2Cl^-(aq) \rightarrow ClO_3^-(aq) + H_2O(l) + Cl_2(g)$

b) 1: $\underset{+7}{Mn}O_4^-(aq) + \underset{0}{Al}(s) \rightarrow \underset{+2}{Mn}^{2+}(aq) + \underset{+3}{Al}^{3+}(aq)$

2: $MnO_4^- \rightarrow Mn^{2+}$ $\qquad Al \rightarrow Al^{3+}$

3: $MnO_4^- + 8H^+ \rightarrow Mn^{2+} + 4H_2O$ $\qquad Al \rightarrow Al^{3+}$

4: $MnO_4^- + 8H^+ + 5e^- \rightarrow Mn^{2+} + 4H_2O$ $\qquad Al \rightarrow Al^{3+} + 3e^-$

5: $3MnO_4^- + 24H^+ + 15e^- \rightarrow 3Mn^{+2} + 12H_2O$ $\qquad 5Al \rightarrow 5Al^{3+} + 15e^-$

6: $3MnO_4^-(aq) + 24H^+(aq) + 5Al(s) \rightarrow 3Mn^{2+}(aq) + 12H_2O(l) + 5Al^{3+}(aq)$

c) 1: $\underset{0}{Br_2}(aq) + \underset{0}{Sn}(s) \rightarrow \underset{+2}{Sn}^{2+}(aq) + \underset{-1}{Br}^-(aq)$

2: $Br_2 \rightarrow Br^-$ $\qquad Sn \rightarrow Sn^{2+}$

3: $Br_2 \rightarrow 2Br^-$ $\qquad Sn \rightarrow Sn^{2+}$

4: $Br_2 + 2e^- \rightarrow 2Br^-$ $\qquad Sn \rightarrow Sn^{2+} + 2e^-$

5: $Br_2 + 2e^- \rightarrow 2Br^-$ $\qquad Sn \rightarrow Sn^{2+} + 2e^-$

6: $Br_2(aq) + Sn(s) \rightarrow Sn^{2+}(aq) + 2Br^-(aq)$

69. a) 1. $\underset{+1\ -2}{ClO}^-(aq) + \underset{+3\ -2\ +1}{Cr(OH)}_4^-(aq) \rightarrow \underset{+6\ -2}{CrO}_4^{2-}(aq) + \underset{-1}{Cl}^-(aq)$

2. $ClO^- \rightarrow Cl^-$ $\qquad Cr(OH)_4^- \rightarrow CrO_4^{2-}$

3. $ClO^- + 2H^+ \rightarrow Cl^- + H_2O$ $\qquad Cr(OH)_4^- \rightarrow CrO_4^{2-} + 4H^+$

4. $ClO^- + 2H^+ + 2e^- \rightarrow Cl^- + H_2O$ $\qquad Cr(OH)_4^- \rightarrow CrO_4^{2-} + 4H^+ + 3e^-$

5. $3ClO^- + 6H^+ + 6e^- \rightarrow 3Cl^- + 3H_2O$ $\qquad 2Cr(OH)_4^- \rightarrow 2CrO_4^{2-} + 8H^+ + 6e^-$

6. $3ClO^- + 2Cr(OH)_4^- \rightarrow 2CrO_4^{2-} + 3Cl^- + 2H^+ + 3H_2O$

7. Overall: $3ClO^- + 2Cr(OH)_4^- + 2OH^- \rightarrow 2CrO_4^{2-} + 3Cl^- + 5H_2O$

b) 1. $\underset{+7\ -2}{MnO}_4^-(aq) + \underset{-1}{Br}^-(aq) \rightarrow \underset{+4\ -2}{MnO}_2(s) + \underset{+5\ -2}{BrO}_3^-(aq)$

2. $MnO_4^- \rightarrow MnO_2$ $\qquad Br^- \rightarrow BrO_3^-$

3. $MnO_4^- + 4H^+ \rightarrow MnO_2 + 2H_2O$ $\qquad Br^- + 3H_2O \rightarrow BrO_3^- + 6H^+$

4. $MnO_4^- + 4H^+ + 3e^- \rightarrow MnO_2 + 2H_2O$ $\qquad Br^- + 3H_2O \rightarrow BrO_3^- + 6H^+ + 6e^-$

5. $2MnO_4^- + 8H^+ + 6e^- \rightarrow 2MnO_2 + 4H_2O$ $\qquad Br^- + 3H_2O \rightarrow BrO_3^- + 6H^+ + 6e^-$

6. $2MnO_4^- + Br^- + 2H^+ \rightarrow 2MnO_2 + BrO_3^- + H_2O$

7. Overall: $2MnO_4^- + Br^- + H_2O \rightarrow 2MnO_2 + BrO_3^- + 2OH^-$

The Activity Series

71. a) Ag -For the elements listed, it is lowest on the activity series of metals.

73. b) Cu^{2+}

75. b) Al

77. a) no reaction
b) spontaneous
c) spontaneous
d) no reaction

79. You could use any of the following metals: Fe, Cr, Zn, Mn, Al, Mg, Na, Ca, K, Li.

81. Mg will reduce Al^{3+} but not Na^{+}.

83. a) no reaction
b) $Fe + 2HCl \rightarrow H_2 + Fe^{2+} + 2Cl^-$
c) no reaction
d) $2Al + 6HCl \rightarrow 3H_2 + 2Al^{3+} + 6Cl^-$

Batteries, Electrochemical Cells, and Electrolysis

85. The electrochemical cell:

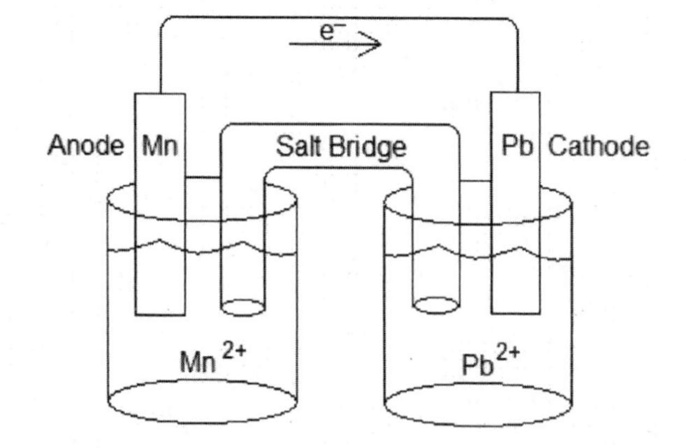

87. Choice d would produce the highest voltage.

89. The overall reaction for an alkaline battery is:

cathode: $2MnO_2(s) + 2H_2O(l) + 2e^- \rightarrow 2MnO(OH)(s) + 2OH^-(aq)$

+anode: $Zn(s) + 2OH^-(aq) \rightarrow Zn(OH)_2(s) + 2e^-$

Overall: $2MnO_2(s) + 2H_2O(l) + Zn(s) \rightarrow 2MnO(OH)(s) + Zn(OH)_2(s)$

91. The electrolysis cell for electroplating copper:

anode: $Cu(s) \rightarrow Cu^{2+}(aq) + 2e^-$

cathode: $Cu^{2+}(aq) + 2e^- \rightarrow Cu(s)$

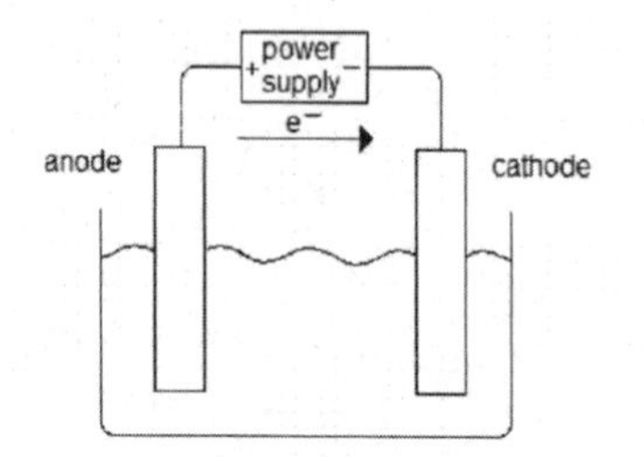

Corrosion

93. a) yes
    b) no
    c) yes

Cumulative Problems

95. a) Redox reaction with Zn being oxidized and Co being reduced.
    b) Not a redox reaction.
    c) Not a redox reaction.
    d) Redox reaction with K being oxidized and Br_2 being reduced.

97. 1: $\underset{+7}{Mn}O_4^-(aq) + \underset{0}{Zn}(s) \rightarrow \underset{+2}{Mn^{2+}}(aq) + \underset{+2}{Zn^{2+}}(aq)$

2: $MnO_4^- \rightarrow Mn^{2+}$ $\qquad Zn \rightarrow Zn^{2+}$

3: $MnO_4^- + 8H^+ \rightarrow Mn^{2+} + 4H_2O$ $\qquad Zn \rightarrow Zn^{2+}$

4: $MnO_4^- + 8H^+ + 5e^- \rightarrow Mn^{2+} + 4H_2O$ $\qquad Zn \rightarrow Zn^{2+} + 2e^-$

5: $2MnO_4^- + 16H^+ + 10e^- \rightarrow 2Mn^{2+} + 8H_2O$ $\qquad 5Zn \rightarrow 5Zn^{2+} + 10e^-$

6: $2MnO_4^-(aq) + 16H^+(aq) + 5Zn(s) \rightarrow 2Mn^{2+}(aq) + 8H_2O(l) + 5Zn^{2+}(aq)$

$$2.85 \text{ g Zn} \times \frac{1 \text{ mol Zn}}{65.39 \text{ g}} \times \frac{2 \text{ mol KMnO}_4}{5 \text{ mol Zn}} \times \frac{1 \text{ L}}{0.500 \text{ mol KMnO}_4} = 0.0349 \text{ L}$$

99. Yes, Mg is more easily oxidized than Ag, therefore Mg will react with Ag^+.

1. $Mg(s) + Ag^+(aq) \rightarrow Mg^{2+}(aq) + Ag(s)$
2. $Mg(s) \rightarrow Mg^{2+}(aq)$ $\qquad Ag^+(aq) \rightarrow Ag(s)$
3. $Mg(s) \rightarrow Mg^{2+}(aq)$ $\qquad Ag^+(aq) \rightarrow Ag(s)$ (no change)
4. $Mg(s) \rightarrow Mg^{2+}(aq) + 2e^-$ $\qquad Ag^+(aq) + e^- \rightarrow Ag(s)$
5. $Mg(s) \rightarrow Mg^{2+}(aq) + 2e^-$ $\qquad 2Ag^+(aq) + 2e^- \rightarrow 2Ag(s)$
6. $Mg(s) + 2Ag^+(aq) \rightarrow Mg^{2+}(aq) + 2Ag(s)$

101. $5H_2O_2(aq) + 2MnO_4^-(aq) + 6H^+(aq) \rightarrow 2Mn^{2+}(aq) + 5O_2(g) + 8H_2O(l)$

$$0.03481 \text{ L MnO}_4^- \times \frac{0.0998 \text{ mol MnO}_4^-}{1 \text{ L}} \times \frac{5 \text{ mol H}_2\text{O}_2}{2 \text{ mol MnO}_4^-} \times \frac{34.01 \text{ g}}{1 \text{ mol H}_2\text{O}_2} = 0.295 \text{ g H}_2\text{O}_2$$

$$\frac{0.295 \text{ g H}_2\text{O}_2}{10.00 \text{ g}} \times 100\% = 2.95\%$$

103. $5.8 \text{ g Ag} \times \frac{1 \text{ mol Ag}}{107.87 \text{ g}} \times \frac{1 \text{ mol e}^-}{1 \text{ mol Ag}} = 0.054 \text{ mol e}^-$

105. a) $6HI + 2Cr \rightarrow 2Cr^{3+} + 3H_2 + 6I^-$

$$5.95 \text{ g Cr} \times \frac{1 \text{ mol Cr}}{52.00 \text{ g}} \times \frac{6 \text{ mol HI}}{2 \text{ mol Cr}} \times \frac{1 \text{ L}}{3.5 \text{ mol HI}} = 0.098 \text{ L}$$

b) $6HI + 2Al \rightarrow 2Al^{3+} + 3H_2 + 6I^-$

$$2.15 \text{ g Al} \times \frac{1 \text{ mol Al}}{26.98 \text{ g}} \times \frac{6 \text{ mol HI}}{2 \text{ mol Al}} \times \frac{1 \text{ L}}{3.5 \text{ mol HI}} = 0.068 \text{ L}$$

c) no reaction

d) no reaction

107. $2Al(s) + 6HCl(aq) \rightarrow 2AlCl_3(aq) + 3H_2(g)$

$$\frac{6.0 \text{ mol HCl}}{1 \text{ L}} \times 0.000050 \text{ L} \times \frac{2 \text{ mol Al}}{6 \text{ mol HCl}} \times \frac{26.98 \text{ g}}{1 \text{ mol Al}} = 0.0027 \text{ g}$$

$$\text{Volume Dissolved:} \frac{0.0027\text{g}}{2.7\text{g/cm}^3} = 1.0x10^{-3}\text{cm}^3$$

$$\text{Volume Cylinder} = \pi r^2 h \Rightarrow 1.0x10^{-3}\text{cm}^3 = 3.14 r^2 (0.0028 \text{ cm})$$

$$r = \sqrt{\frac{1.0 \times 10^{-3}\text{cm}^3}{3.14(0.0028 \text{ cm})}} = 0.337 \text{ cm}$$

diameter = 0.67cm

109. $$1.0 \text{ g Ag} \times \frac{1 \text{ mol Ag}}{107.87 \text{ g}} \times \frac{1 \text{ mol e}^-}{1 \text{ mol Ag}} \times \frac{6.022 \times 10^{23} \text{e}^-}{1 \text{ mol e}^-} \times \frac{1.60 \times 10^{-19}\text{C}}{1 \text{ e}^-} \times \frac{1 \text{ s}}{0.100\text{C}} = 8.9 \times 10^3 \text{s}$$

Highlight Problems

111. The sketch should show aluminum atoms going into solution as +3 ions, which dissolve the metal electrode. The copper ions are forming solid, elemental copper atoms on the aluminum strip.

113. The sketch should show the formation of an increased number of zinc ions in solution and the loss of zinc atoms from the surface of the anode. The cathode should have an increased number of nickel atoms and a corresponding decrease in the nickel ions from solution.

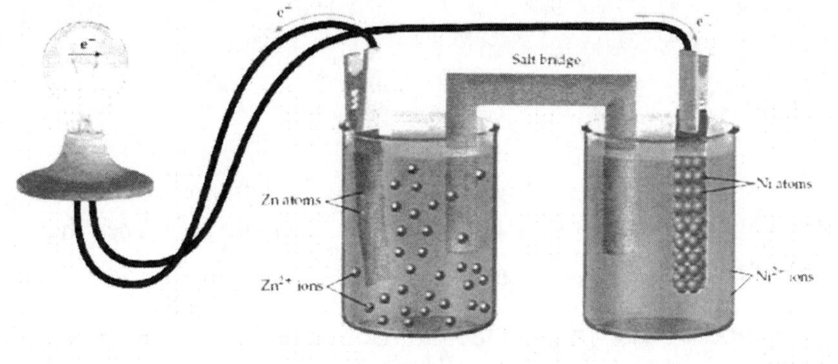

Radioactivity and Nuclear Chemistry

17

Questions

1. Radioactivity is the emission of tiny, invisible (to the human eye) particles by the nuclei of certain atoms. A radioactive atom will spontaneously emit these tiny, invisible particles.

3. Uranic rays were the name given to the radioactive particles being emitted from uranium minerals in Becquerel's studies.

5. X = chemical symbol, used to identify element
   Z = atomic number, the number of protons in the nucleus; determines the identity of the element
   A = atomic mass, the sum of the number of protons and the number of neutrons in the nucleus

7. Alpha (α) radiation occurs when the nucleus emits a particle that contains 2 protons and 2 neutrons. Alpha particles are symbolized by ${}^{4}_{2}He$.

9. Alpha particles have a high ionizing power in comparison with other radioactive particles; however, they have very low penetrating power in comparison with the other radioactive particles.

11. When an atom emits a beta particle, a neutron in the nucleus is converted to a proton. As a result, the atomic number increases by 1 while the atomic mass remains constant.

13. Gamma radiation is a type of electromagnetic radiation that is essentially a high energy photon. The gamma particle is symbolized by ${}^{0}_{0}\gamma$.

15. Gamma rays have low ionizing power and high penetrating power in comparison with other radioactive particles.

17. When an atom emits a positron, it converts a proton to a neutron, which converts it into an atom of the next lighter element (Z decreases by 1, while A remains constant).

19. A nuclear equation represents nuclear processes such as radioactivity. For a nuclear equation to be balanced, the sum of the atomic numbers on both sides of the equation must be equal, and the sum of the mass numbers on both sides of the equation must be equal.

21. A film-badge dosimeter is a radioactivity detector used as a safety precaution for people who work with or near radioactive compounds. The dosimeter is a small piece of photographic film that is attached to clothing and regularly collected and developed in order to monitor the amount of radioactivity to which the badge has been exposed in the recent past.

23. In a scintillation counter, the radioactive particles pass through a crystal of NaI or CsI, which emits UV-Vis photons as it becomes excited by the radioactive particles. The photons are then detected and converted into an electrical signal.

25. The half-life of a radioactive nuclide is the time it will take for ½ of the original parent nuclides to undergo decay. The half-life can be used to determine radioactive decay rates.

27. Radon in the environment comes from the radioactive decay series of uranium. It presents a danger because radon is a gaseous compound and its daughter nuclides can attach to dust particles, which can be inhaled into the human body and increase a person's risk for developing lung cancer.

29. When an organism dies, it no longer uptakes carbon-14; therefore, the amount of carbon-14 starts to decay at a rate equal to the half-life of carbon-14. By measuring how much carbon-14 remains in an organism, you can use the half-life to determine how long it took the carbon-14 to decay to the current level.

31. ${}^{235}_{92}\mathrm{Ur} + {}^{1}_{0}\mathrm{n} \rightarrow {}^{140}_{56}\mathrm{Ba} + {}^{93}_{36}\mathrm{Kr} + 3{}^{1}_{0}\mathrm{n} + \text{energy}$
Enrico Fermi bombarded uranium-235 with neutrons in the hope that it would undergo beta decay and produce a new element with atomic number 93. Instead, it underwent nuclear fission, breaking into several smaller elements.

33. The nuclear fission reaction is triggered by a neutron colliding with a fissionable nuclide, which produces lighter elements and more neutrons. The neutrons produced in the reaction can then collide with other nuclides producing a chain reaction. This reaction can then be used to make a bomb, because a tremendous amount of energy is released during each fission reaction.

35. The main goal of the Manhattan Project was to build an atomic bomb before the Germans did. The project was led by J.R. Oppenheimer.

37. The control rods serve to control the rate of the nuclear fission reaction in the reactor core so that the temperature is maintained at a desired level. If the reactor core temperature is too low, the control rods are raised which allows more neutrons to collide with uranium, producing heat and the fission products. If the reactor core temperature is too hot, the control rods are lowered into the core, absorbing a greater number of neutrons, therefore fewer neutrons will collide with uranium and less heat is produced.

39. A nuclear power plant cannot detonate like a nuclear bomb because the uranium fuel is not enriched with the quantity of uranium-235 that is needed to produce a bomb.

41. The traditional nuclear bombs are of the fission type. However, hydrogen bombs are based on a fusion reaction. It should be noted that in order for a hydrogen bomb to work, a small fission reaction takes place to generate sufficient heat to initiate the fusion reaction.

43. Radiation can ionize molecules in living organisms.

45. Radiation can increase the risk of cancer because it can damage DNA, which can cause cells to grow abnormally.

47. The main unit of radiation is the roentgen equivalent man, aka rem. A typical person is exposed to one-third of a rem per year.

49. Isotope scanning is a method used to diagnose different diseases based on the use of radioactive isotopes and their ability to target different organs and tissues. The radioactive isotopes are detected with either photographic film or a scintillation counter.

Problems

Isotopic and Nuclear Particle Symbols

51. ${}^{210}_{82}\mathrm{Pb}$

53. Protons (Z): 81
Neutrons (A–Z): 126

55. a) beta particle
b) neutron
c) gamma ray

57.

Chemical Symbol	Atomic Number (Z)	Mass Number (A)	#Protons	#Neutrons
Tc	43	95	43	52
Ba	56	128	56	72
Eu	63	145	63	82
Fr	87	223	87	136

Radioactive Decay

59. a) $^{234}_{92}\text{U} \rightarrow {}^{230}_{90}\text{Th} + {}^{4}_{2}\text{He}$

b) $^{230}_{90}\text{Th} \rightarrow {}^{226}_{88}\text{Ra} + {}^{4}_{2}\text{He}$

c) $^{226}_{88}\text{Ra} \rightarrow {}^{222}_{86}\text{Rn} + {}^{4}_{2}\text{He}$

d) $^{222}_{86}\text{Rn} \rightarrow {}^{218}_{84}\text{Po} + {}^{4}_{2}\text{He}$

61. a) $^{214}_{82}\text{Pb} \rightarrow {}^{214}_{83}\text{Bi} + {}^{0}_{-1}\text{e}$

b) $^{214}_{83}\text{Bi} \rightarrow {}^{214}_{84}\text{Po} + {}^{0}_{-1}\text{e}$

c) $^{231}_{90}\text{Th} \rightarrow {}^{231}_{91}\text{Pa} + {}^{0}_{-1}\text{e}$

d) $^{227}_{89}\text{Ac} \rightarrow {}^{227}_{90}\text{Th} + {}^{0}_{-1}\text{e}$

63. a) $^{11}_{6}\text{C} \rightarrow {}^{11}_{5}\text{B} + {}^{0}_{+1}\text{e}$

b) $^{13}_{7}\text{N} \rightarrow {}^{13}_{6}\text{C} + {}^{0}_{+1}\text{e}$

c) $^{15}_{8}\text{O} \rightarrow {}^{15}_{7}\text{N} + {}^{0}_{+1}\text{e}$

65. $^{241}_{94}\text{Pu} \rightarrow {}^{241}_{95}\text{Am} + {}^{0}_{-1}\text{e}$

$^{241}_{95}\text{Am} \rightarrow {}^{237}_{93}\text{Np} + {}^{4}_{2}\text{He}$

$^{237}_{93}\text{Np} \rightarrow {}^{233}_{91}\text{Pa} + {}^{4}_{2}\text{He}$

$^{233}_{91}\text{Pa} \rightarrow {}^{233}_{92}\text{U} + {}^{0}_{-1}\text{e}$

67. $^{232}_{90}\text{Th} \rightarrow {}^{228}_{88}\text{Ra} + {}^{4}_{2}\text{He}$

$^{228}_{88}\text{Ra} \rightarrow {}^{228}_{89}\text{Ac} + {}^{0}_{-1}\text{e}$

$^{228}_{89}\text{Ac} \rightarrow {}^{228}_{90}\text{Th} + {}^{0}_{-1}\text{e}$

$^{228}_{90}\text{Th} \rightarrow {}^{224}_{88}\text{Ra} + {}^{4}_{2}\text{He}$

Half-Life

69. $100{,}000 \xrightarrow{2\text{ days}} 50{,}000 \xrightarrow{2\text{ days}} 25{,}000 \xrightarrow{2\text{ days}} 12{,}500 \xrightarrow{2\text{ days}} 6{,}250 \xrightarrow{2\text{ days}} 3{,}125$ (10 days)

There would be 3,125 radioactive atoms remaining after 10 days.

71. $5.0\times10^{-2} \xrightarrow{6\text{ hrs}} 2.5\times10^{-2} \xrightarrow{6\text{ hrs}} 1.25\times10^{-2} \xrightarrow{6\text{ hrs}} 6.25\times10^{-3}$
It would take 18 hrs for technetium-99 to decay to 6.3×10^{-3} mg.

73. $2.80 \xrightarrow{1\text{st}} 1.40 \xrightarrow{2\text{nd}} 0.700 \xrightarrow{3\text{rd}} 0.350 \xrightarrow{4\text{th}} 0.175 \xrightarrow{5\text{th}} 0.0875$
It would take 5 half lives or 1.22×10^{6} years.

75. $2.45 \xrightarrow{3.8\text{ days}} 1.23 \xrightarrow{3.8\text{ days}} 0.613 \xrightarrow{3.8\text{ days}} 0.306$
There would be 0.306 grams of the isotope after 11.4 days.

77. Ga-67 > P-32 > Cr-51 > Sr-89

Radiocarbon Dating

79. The age of the boat would be equal to one half-life of C-14, or approximately 5,730 years old.

81. The skull has undergone six half-lives of C-14, which would make it approximately 34,380 years old.

Fission and Fusion

83. ${}^{235}_{92}\text{U} + {}^{1}_{0}\text{n} \rightarrow {}^{144}_{54}\text{Xe} + {}^{90}_{38}\text{Sr} + 2{}^{1}_{0}\text{n}$; 2 neutrons produced.

85. ${}^{2}_{1}\text{H} + {}^{2}_{1}\text{H} \rightarrow {}^{3}_{2}\text{He} + {}^{1}_{0}\text{n}$

Cumulative Problems

87. a) ${}^{1}_{1}\text{p} + {}^{9}_{4}\text{Be} \rightarrow \underline{{}^{6}_{3}\text{Li}} + {}^{4}_{2}\text{He}$
b) ${}^{209}_{83}\text{Bi} + \underline{{}^{64}_{28}\text{Ni}} \rightarrow {}^{272}_{111}\text{Rg} + {}^{1}_{0}\text{n}$
c) ${}^{179}_{74}\text{W} + {}^{0}_{-1}\text{e}^{-} \rightarrow \underline{{}^{179}_{73}\text{Ta}}$

89. ${}^{238}_{92}\text{U} + {}^{1}_{0}\text{n} \rightarrow {}^{239}_{92}\text{U}$; ${}^{239}_{92}\text{U} \rightarrow {}^{0}_{-1}\beta + {}^{239}_{93}\text{Np}$; ${}^{239}_{93}\text{Np} \rightarrow {}^{0}_{-1}\beta + {}^{239}_{94}\text{Pu}$

91. $\dfrac{3.2x10^{-11}\text{ J}}{\text{atom}} \times \dfrac{6.022\times10^{23}\text{ atoms}}{1\text{ mol U}} = 1.9\times10^{13}\text{ J/mol}$

$\dfrac{1.9\times10^{13}\text{ J}}{\text{mol}} \times \dfrac{1\text{ mol U}}{238\text{ g}} \times \dfrac{1000\text{ g}}{1\text{ kg}} = 8.1\times10^{13}\text{ J/kg}$

93. In one half-life (5 days), we will lose 0.60 g of material. Assuming that each atom produces one beta particle in the decay process, the number of beta emissions is

$$0.60\text{ g Bi} \times \frac{1\text{ mol Bi}}{208.98\text{ g}} \times \frac{6.022\times10^{23}\text{ atoms}}{1\text{ mol Bi}} \times \frac{1\ \beta\text{ particle}}{1\text{ atom}} = 1.7\times10^{21}\beta\text{ particles}$$

95. $\dfrac{0.400\text{ rem}}{0.585\text{ rem}} \times 100\% = 68.4\%$ due to radon

97. $\dfrac{1.6\times10^{3}\text{ yrs}}{\text{half-life}} \times \dfrac{365\text{ d}}{1\text{ yr}} = 5.84\times10^{5}\text{ days/half-life}$

$$45\text{ days} \times \frac{\text{half-life}}{5.84\times10^{5}\text{ days}} = 7.7\times10^{-5}\text{ half-lives}$$

$$\underbrace{(0.5)^{7.7\times10^{-5}} = 0.999946629}_{\text{Fraction remaining}}$$

$$\underbrace{1.5\text{ g Rn} \times 0.999946629 = 1.499919944}_{\text{Mass remaining}}$$

$$8.01\times10^{-5}\text{g Ra} \times \frac{1\text{ mol Ra}}{226.03\text{ g}} \times \frac{1\text{ mol Rn}}{1\text{mol Ra}} = 3.54\times10^{-7}\text{mol Rn}$$

$$V = \frac{nRT}{P} = \frac{(3.54\times10^{-7}\text{mol Rn})(0.08206\text{ L}\bullet\text{atm/mol}\bullet\text{K})(298.15\text{ K})}{1\text{ atm}} = 8.67\times10^{-6}\text{L Rn}$$

Highlight Problems

99. The missing nucleus contains 9 protons and 7 neutrons (fluorine-16).

101. The missing nucleus contains 5 protons and 5 neutrons (boron-10).

Organic Chemistry

Questions

1. Organic molecules are responsible for most odors.

3. At the end of the eighteenth century, it was believed that organic compounds came from living things and were easily decomposed, while inorganic compounds came from the earth and were more difficult to decompose. A final difference is that many inorganic compounds could be easily synthesized, but organic compounds could not be.

5. Carbon chemistry is complex because of carbon's ability to bond with itself to form chains, branches, and ring structures. As a result, the number of molecules that can be formed from carbon is very large. It is this complexity and diversity of carbon-based compounds that makes life possible.

7. Hydrocarbons are compounds that contain only carbon and hydrogen atoms. The main uses of hydrocarbons are as fuels and as raw materials in the synthesis of many products, including fabrics, soaps, dyes, cosmetics, drugs, plastic, and rubber.

9. Alkanes are considered saturated hydrocarbons because they contain only single bonds and they have the maximum number of hydrogen atoms possible (i.e., saturated). Alkenes and alkynes are considered unsaturated hydrocarbons because they contain double and triple bonds, and therefore have fewer hydrogen atoms than the similar alkane compound.

11. The n-alkane compounds have all carbon atoms bonded in a straight chain, where the branched alkanes have branches of carbon atoms coming off of a main straight chain of carbon atoms.

13. Alkenes are hydrocarbons that contain at least one double bond between carbon atoms, where alkanes only contain single bonds.

15. Hydrocarbon combustion reactions are the burning of hydrocarbons in the presence of oxygen. An example of a hydrocarbon combustion reaction is:

$$CH_3CH_2CH_3(g) + 5O_2(g) \rightarrow 3CO_2(g) + 4H_2O(g)$$

17. Alkene addition reactions occur when atoms add across the double bond. An example of an alkene addition reaction is:

$$CH_2{=}CH_2(g) + Cl_2(g) \rightarrow CH_2ClCH_2Cl(g)$$

19. The structure of benzene is 6 carbon atoms connected together in a ring, with a single hydrogen atom bonded to each carbon. Two common ways of showing the structure of benzene are shown below.

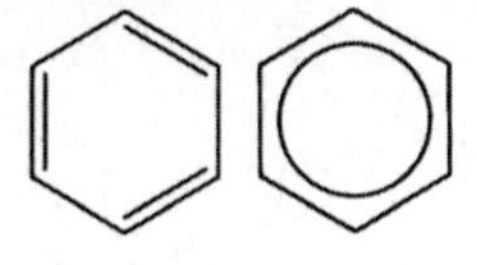

21. The generic structure of alcohols is ROH. Two examples of alcohols are ethanol: CH_3CH_2OH and 1-butanol: $CH_3CH_2CH_2CH_2OH$.

23. The generic structure of ethers is R-O-R. Two examples of ethers are dimethyl ether: CH_3OCH_3 and diethyl ether: $CH_3CH_2OCH_2CH_3$.

25.

Functional Group	Generic Structure	Example
aldehydes	$R-\overset{O}{\overset{\|}{C}}-H$	$CH_3-\overset{O}{\overset{\|}{C}}-H$ (acetaldehyde)
ketones	$R-\overset{O}{\overset{\|}{C}}-R$	$CH_3-\overset{O}{\overset{\|}{C}}-CH_3$ (acetone)

27.

Functional Group	Generic Structure	Example
carboxylic acid	$R-\overset{O}{\overset{\|}{C}}-OH$	$CH_3-\overset{O}{\overset{\|}{C}}-OH$ (acetic acid)
esters	$R-\overset{O}{\overset{\|}{C}}-OR$	$CH_3-CH_2-\overset{O}{\overset{\|}{C}}-O-CH_2-CH_3$ (ethyl propanoate)

29. The generic structure of amines is NR_3. Examples of specific amines are ammonia $H-\overset{H}{\overset{|}{N}}-H$ and ethylamine $CH_3CH_2-\overset{H}{\overset{|}{N}}-H$

31. A polymer is a long, chainlike molecule that is made up of small repeating units called monomers. A polymer is made up of one type of monomer, where a copolymer has two different monomers.

Problems

Hydrocarbons

33. Choices c and d are hydrocarbons because they contain only C and H.

35. a) alkyne (2n-2)
    b) alkane (2n+2)
    c) alkyne (2n-2)
    d) alkene (2n)

Alkanes

37. a) $CH_3CH_2CH_2CH_2CH_2CH_2CH_3$

```
  H H H H H H H
H-C-C-C-C-C-C-C-H
  H H H H H H H
```

b) $CH_3CH_2CH_2CH_2CH_2CH_2CH_2CH_3$

```
  H H H H H H H H
H-C-C-C-C-C-C-C-C-H
  H H H H H H H H
```

c) $CH_3CH_2CH_2CH_2CH_2CH_3$

```
  H H H H H H
H-C-C-C-C-C-C-H
  H H H H H H
```

d) CH_3CH_3

```
  H H
H-C-C-H
  H H
```

```
  H H H H H
H-C-C-C-C-C-H
  H H H H H
```

39. Two isomers of butane:

```
      CH3
       |
H3C—CH—CH3      H3C—CH2—CH2—CH3
```

41. Five isomers of octane (18 total isomers):

```
  H H H H H H H H          H CH3H H H H H            H H CH3H H H H
  | | | | | | | |          | |  | | | | |            | | |  | | | |
H-C-C-C-C-C-C-C-C-H      H-C-C--C-C-C-C-C-H        H-C-C-C--C-C-C-C-H
  | | | | | | | |          | |  | | | | |            | | |  | | | |
  H H H H H H H H          H H  H H H H H            H H H  H H H H

  H H H CH3H H H           H CH3H H H H              H CH3H H H H
  | | | |  | | |           | |  | | | |              | |  | | | |
H-C-C-C-C--C-C-C-H       H-C-C--C-C-C-C-H          H-C-C--C-C-C-C-H
  | | | |  | | |           | |  | | | |              | |  | | | |
  H H H H  H H H           H CH3H H H H              H H CH3H H H

  H CH3H H H H             H CH3H H H H              H H CH3 H H H
  | |  | | | |             | |  | | | |              | | |   | | |
H-C-C--C-C-C-C-H         H-C-C--C-C-C-C-H          H-C-C-C---C-C-C-H
  | |  | | | |             | |  | | | |              | | |   | | |
  H H  H CH3H H            H H  H H CH3H             H H CH3 H H H

  H H CH3H H H             H CH3H H H                H CH3H H H
  | | |  | | |             | |  | | |                | |  | | |
H-C-C-C--C-C-C-H         H-C-C--C-C-C-H            H-C-C--C-C-C-H
  | | |  | | |             | |  | | |                | |  | | |
  H H H CH3H H             H CH3CH3H H               H CH3H CH3H

  H H CH3H H               H H H H H                 H CH3 CH3H
  | | |  | |               | | | | |                 | |   |  |
H-C-C-C--C-C-H           H-C-C-C-C-C-H             H-C-C---C--C-H
  | | |  | |               | | | | |                 | |   |  |
  H CH3CH3H H              H CH3CH3 CH3 H            H CH3 CH3H

  H H H H H H              H H H H H                 H H CH3H H
  | | | | | |              | | | | |                 | | |  | |
H-C-C-C-C-C-C-H          H-C-C-C-C-C-H             H-C-C-C--C-C-H
  | | | | | |              | | | | |                 | | |  | |
  H H CH2H H H             H CH3CH2H H               H H CH2H H
      |                           |                      |
      CH3                         CH3                    CH3
```

43. a) n-pentane
    b) 2-methylbutane
    c) 4-ethyl-2-methylhexane
    d) 3,3-dimethylpentane

45. Alkane structures:

a)
```
         CH3
         |
H3C—— CH—CH2—CH3
```

b)
```
              CH3
              |
         CH3  CH2
         |    |
H3C—CH—CH2—CH2—CH2—CH3
```

c)
```
     H3C—CH—CH3
          |
H3C—CH2—CH2—CH2—CH2—CH3
```

d)
```
H3C—CH—CH2—CH2—CH—CH2—CH2—CH3
     |              |
     CH3            CH3
```

47. a) n-pentane
    b) 3-methylhexane
    c) 2,3-dimethylpentane

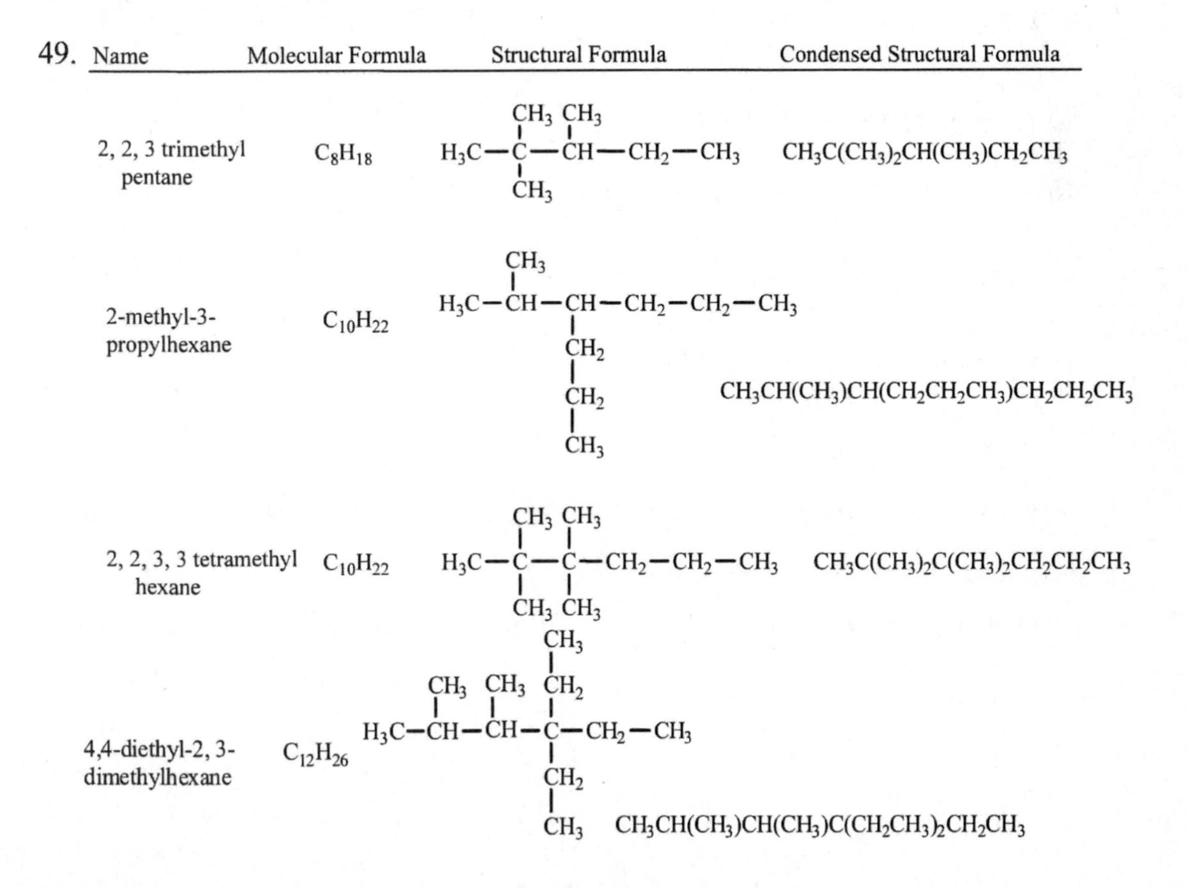

49.

Name	Molecular Formula	Structural Formula	Condensed Structural Formula
2, 2, 3 trimethyl pentane	C_8H_{18}	$H_3C-C(CH_3)_2-CH(CH_3)-CH_2-CH_3$	$CH_3C(CH_3)_2CH(CH_3)CH_2CH_3$
2-methyl-3-propylhexane	$C_{10}H_{22}$	$H_3C-CH(CH_3)-CH(CH_2-CH_2-CH_3)-CH_2-CH_2-CH_3$	$CH_3CH(CH_3)CH(CH_2CH_2CH_3)CH_2CH_2CH_3$
2, 2, 3, 3 tetramethyl hexane	$C_{10}H_{22}$	$H_3C-C(CH_3)_2-C(CH_3)_2-CH_2-CH_2-CH_3$	$CH_3C(CH_3)_2C(CH_3)_2CH_2CH_2CH_3$
4,4-diethyl-2, 3-dimethylhexane	$C_{12}H_{26}$	$H_3C-CH(CH_3)-CH(CH_3)-C(CH_2CH_3)_2-CH_2-CH_3$	$CH_3CH(CH_3)CH(CH_3)C(CH_2CH_3)_2CH_2CH_3$

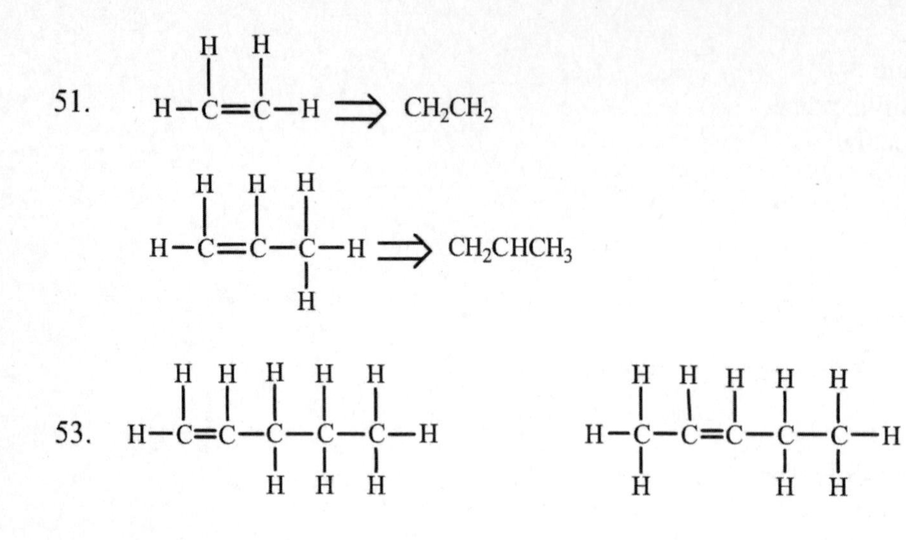

55. a) 2-pentene
    b) 4-methyl-2-pentene
    c) 3,3-dimethyl-1-butene
    d) 3,4-dimethyl-1-hexene

57. a) 2-butyne
    b) 4-methyl-2-pentyne
    c) 4,4-dimethyl-2-hexyne
    d) 3-ethyl-3-methyl-1-pentyne

59. a) $H_3C-CH=CH_3-CH_2-CH_2-CH_3$

    b) $H_3C-CH_2-C\equiv C-CH_2-CH_2-CH_3$

    c) $HC\equiv C-CH(CH_3)-CH_2-CH_3$

    d) $H_3C-CH=CH-C(CH_3)_2-CH_2-CH_3$

61. 1-pentene $H_2C=CH-CH_2-CH_2-CH_3$

    3-methyl-1-butene $H_2C=CH-CH(CH_3)-CH_3$

    2-pentene $H_3C-CH=CH-CH_2-CH_3$

    2-methyl-1butene $H_2C=C(CH_3)-CH_2-CH_3$

    2-methyl-2-butene $H_3C-CH=C(CH_3)-CH_3$

63.

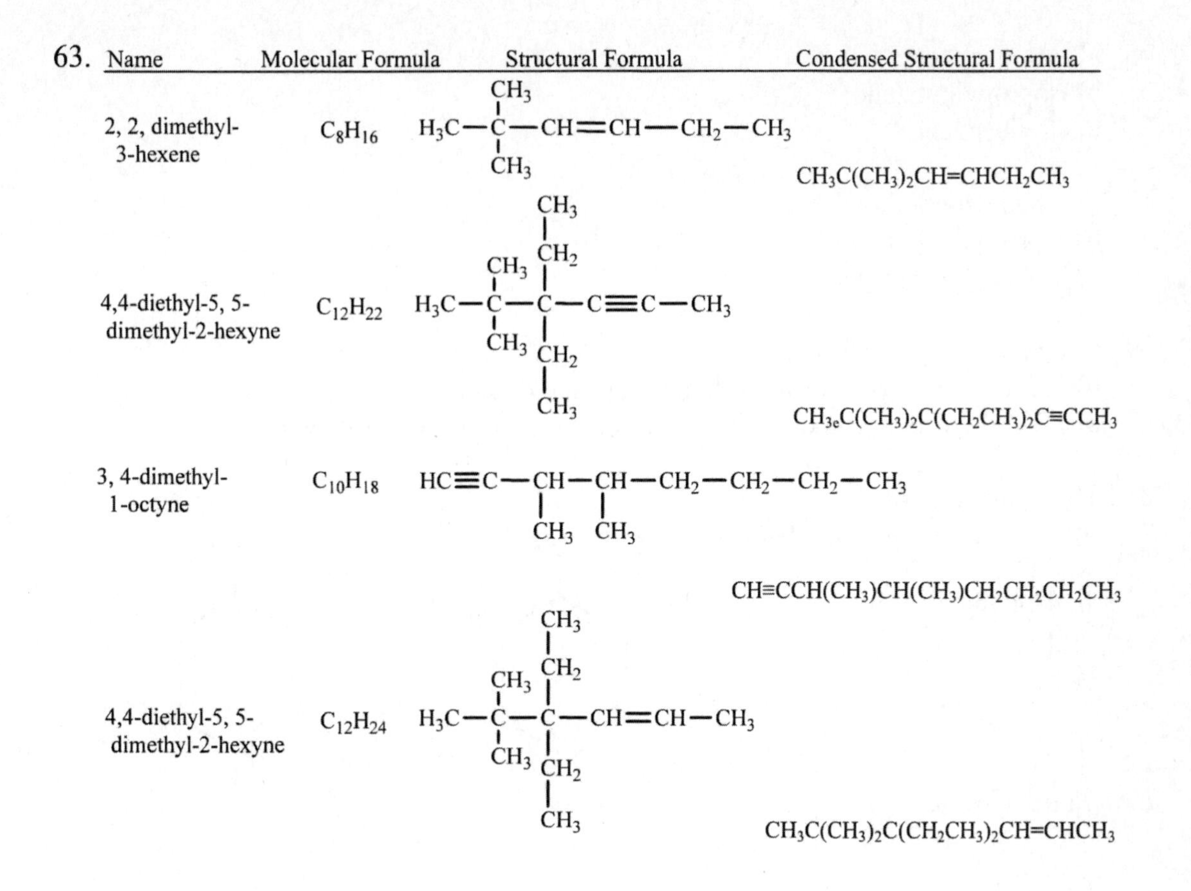

Name	Molecular Formula	Structural Formula	Condensed Structural Formula
2, 2, dimethyl-3-hexene	C_8H_{16}	$H_3C-C(CH_3)(CH_3)-CH=CH-CH_2-CH_3$	$CH_3C(CH_3)_2CH=CHCH_2CH_3$
4,4-diethyl-5, 5-dimethyl-2-hexyne	$C_{12}H_{22}$	$H_3C-C(CH_3)(CH_3)-C(CH_2CH_3)(CH_2CH_3)-C\equiv C-CH_3$	$CH_{3e}C(CH_3)_2C(CH_2CH_3)_2C\equiv CCH_3$
3, 4-dimethyl-1-octyne	$C_{10}H_{18}$	$HC\equiv C-CH(CH_3)-CH(CH_3)-CH_2-CH_2-CH_2-CH_3$	$CH\equiv CCH(CH_3)CH(CH_3)CH_2CH_2CH_2CH_3$
4,4-diethyl-5, 5-dimethyl-2-hexyne	$C_{12}H_{24}$	$H_3C-C(CH_3)(CH_3)-C(CH_2CH_3)(CH_2CH_3)-CH=CH-CH_3$	$CH_3C(CH_3)_2C(CH_2CH_3)_2CH=CHCH_3$

Hydrocarbon Reactions

65. a) $2CH_3CH_3(g) + 7O_2(g) \rightarrow 4CO_2(g) + 6H_2O(g)$
    b) $2CH_2{=}CHCH_3(g) + 9O_2(g) \rightarrow 6CO_2(g) + 6H_2O(g)$
    c) $2CH\equiv CH(g) + 5O_2(g) \rightarrow 4CO_2(g) + 2H_2O(g)$

67. $CH_4(g) + Br_2(g) \rightarrow CH_3Br(g) + HBr(g)$

69. $CH_3CH{=}CHCH_3(g) + Cl_2(g) \rightarrow CH_3CHClCHClCH_3(g)$

71. $CH_2{=}CH_2(g) + H_2(g) \rightarrow CH_3CH_3(g)$

Aromatic Hydrocarbons

73. The structural formula that represents both shorthand formulas is:

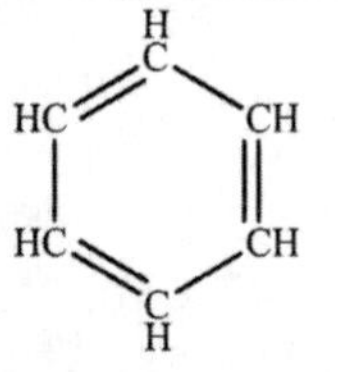

75. a) fluorobenzene
    b) isopropylbenzene
    c) ethylbenzene

77. a) 4-phenyloctane
    b) 5-phenyl-3-heptene
    c) 7-phenyl-2-heptyne

79. a) 1-bromo-2-chlorobenzene
    b) 1,2-diethylbenzene or orthodiethylbenzene or o-diethylbenzene
    c) 1,3-difluorobenzene or metadichlorobenzene or m-dichlorobenzene

81. The structures:

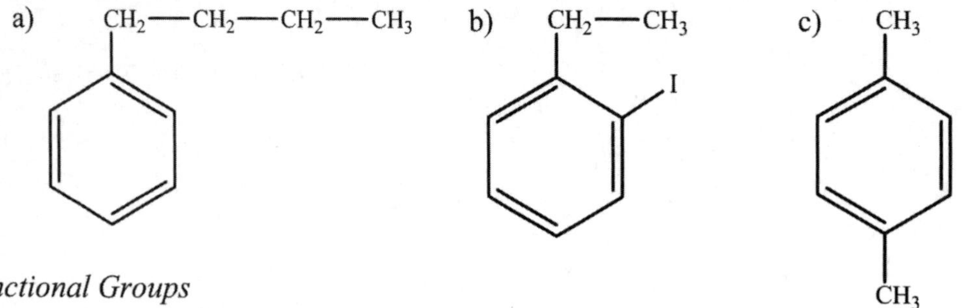

Functional Groups

83.
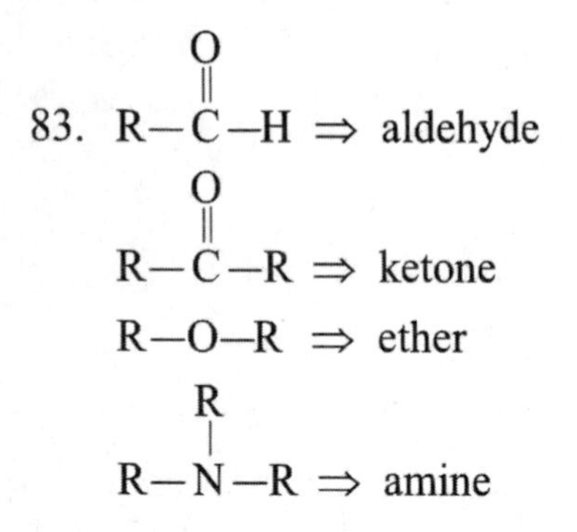

85. Functional Groups and Families:

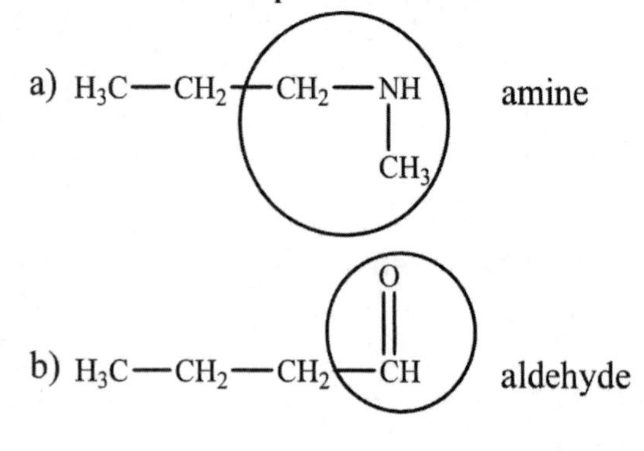

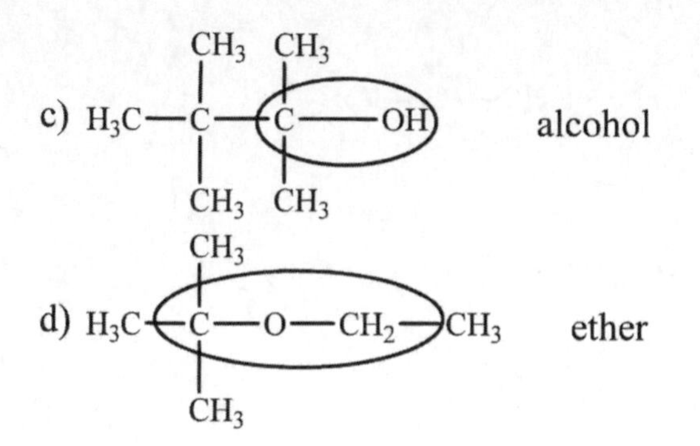

Alcohols

87. a) 2-butanol
    b) 2-methyl-1-propanol
    c) 3-ethyl-1-hexanol
    d) 3-methyl-3-pentanol

89. a) $CH_3CH_2CHCH_2CH_3$
    $\quad\quad\quad\quad |$
    $\quad\quad\quad\quad OH$

    b) $CH_2CH_2CHCH_3$
    $\;|\quad\quad\;\; |$
    $OH\quad\; CH_3$

    c) $CH_3CHCHCH_2CH_2CH_3$
    $\quad\quad |\;\; |$
    $\quad\quad OHCH_2CH_3$

    d) CH_3CH_2OH

Ethers

91. a) $CH_3CH_2CH_2CH_2$-O-$CH_2CH_2CH_2CH_3$
    b) ethyl propyl ether
    c) dipropyl ether
    d) CH_3-O-$CH_2CH_2CH_2CH_2CH_3$

93. a) $H_3C—CH_2—CH_2—CH_2—CH_2—CH_2—CH_2—\overset{\overset{\displaystyle O}{\|}}{C}H_3$
    b) butanal
    c) 4-heptanone

95. a) $H_3C-CH_2-CH_2-CH_2-CH_2-CH_2-CH_2-\overset{\overset{\displaystyle O}{\|}}{C}-OH$

b) methyl ethanoate

c) $H_3C-CH_2-CH_2-\overset{\overset{\displaystyle O}{\|}}{C}-O-CH_2-CH_3$

d) heptanoic acid

Amines

97. a) $H_3C-CH_2-NH-CH_2-CH_3$

b) triethylamine

c) butylpropylamine

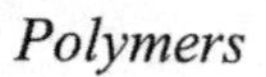

Polymers

99.

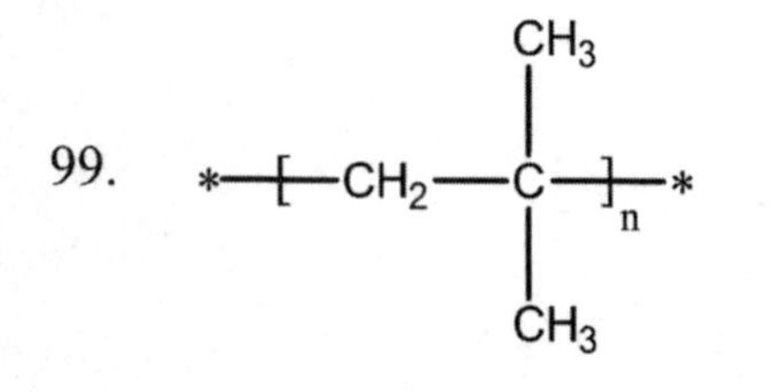

101.

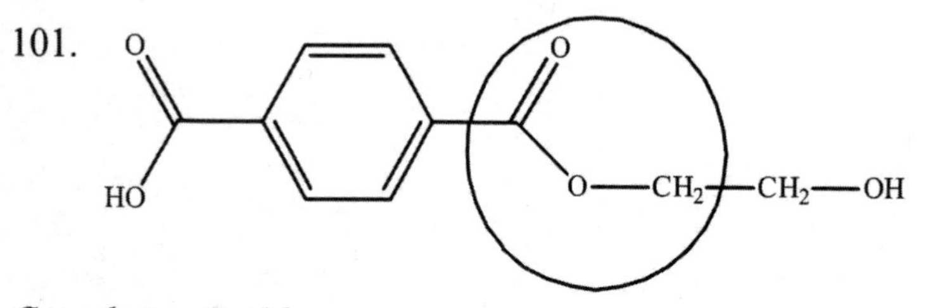

Cumulative Problems

103. a) alcohol
b) amine
c) alkane
d) carboxylic acid
e) ether
f) alkene

105. a) 3-methyl-4-tert-butylheptane
b) 3-methyl butanal
c) 4-isopropyl-3-methyl-2-heptene
d) propyl butanoate

107. a) same molecule
b) isomers
c) same molecule

109. $CH_2{=}CH_2 + HCl \rightarrow CH_3CH_2Cl$

111. $CH_3CH{=}CHCH_3 + H_2 \rightarrow CH_3CH_2CH_2CH_3$

$$15.5\text{ kg 2-butene} \times \frac{1\times10^3\text{g}}{1\text{ kg}} \times \frac{1\text{ mol 2-butene}}{56.12\text{ g}} \times \frac{1\text{ mol }H_2}{1\text{ mol 2-butene}} \times \frac{2.02\text{ g}}{1\text{ mol }H_2}$$

$$= 558\text{ g }H_2$$

113. $2C_8H_{18} + 25O_2 \rightarrow 16CO_2 + 18H_2O$

$$18.9\times10^3\text{g }C_8H_{18} \times \frac{1\text{ mol }C_8H_{18}}{114.2\text{ g}} \times \frac{25\text{ mol }O_2}{2\text{ mol }C_8H_{18}} \times \frac{22.4\text{ L}}{1\text{ mol }O_2} = 4.63\times10^4\text{L }O_2$$

Highlights

115. a) alcohol
b) amine
c) carboxylic acid
d) ester
e) alkane
f) ether

Biochemistry

Questions

1. The goal of the Human Genome Project was to map all of the genetic material (genome) of a human being. One of the surprises from the Human Genome Project results was that humans have only 20,000 – 25,000 genes, and scientists had always predicted that there would be many more.

3. The cell is the smallest structural unit of living things that has the properties associated with life. The main chemical components of the cell can be divided into four classes: carbohydrates, lipids, proteins, and nucleic acids.

5. Glucose is soluble in water because of the many OH groups that allow water to hydrogen bond to glucose. This is important because glucose needs to be transported by blood to the cells and then through the cell wall into the aqueous interior of the cell.

7. During digestion the links in disaccharides and polysaccharides are broken, allowing individual monosaccharides to pass through the intestinal wall and enter the bloodstream.

9. Starch and cellulose are both polysaccharides but the difference is the bond angles that form between saccharide units. Because of the difference in bond angle, humans can digest starch and use it for energy, while cellulose cannot be digested and passes directly through humans.

11. A fatty acid is a carboxylic acid with a long hydrocarbon chain. The general structure of a fatty acid is shown below where R is a carbon chain ranging from 3 to 19 atoms long.

$$R-\overset{\overset{\displaystyle O}{\|}}{C}-OH$$

Generic fatty acid structure

13. Triglycerides are tri-esters composed of glycerol and three fatty acids. Oils and fats are triglycerides.

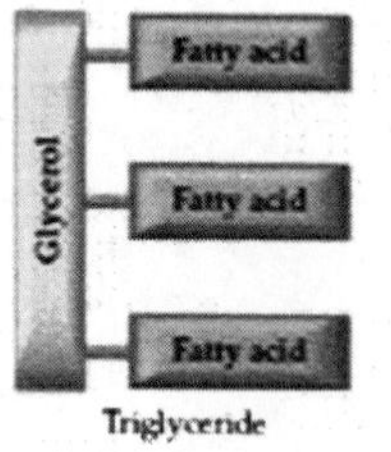

15. Phospholipids are molecules having the same generic structure as triglycerides in which one of the fatty acid esters has been replaced with a phosphate ester. Glycolipids have similar structures and properties as phospholipids, but the polar phosphate group has been replaced by a sugar molecule such as glucose.

17. Steroids are lipids containing a 4-ring structure. Examples include cholesterol, testosterone, and estrogen. Cholesterol is a steroid that is part of cell membranes and also serves as a starting material for the body to synthesize other steroids. Also, steroids serve as male and female hormones in the body.

19. Proteins have many roles in the body such as catalysts, structural unit of muscle, skin and cartilage, transportation of oxygen, disease-fighting antibodies, and as hormones.

21. Amino acids differ from each other by the nature of the R group.

23. The formation of a peptide bond:

$$H_2N-\underset{\displaystyle R}{\underset{|}{CH}}-\overset{\displaystyle O}{\overset{\|}{C}}-OH + H_2N-\underset{\displaystyle R}{\underset{|}{CH}}-\overset{\displaystyle O}{\overset{\|}{C}}-OH \rightarrow H_2N-\underset{\displaystyle R}{\underset{|}{CH}}-\overset{\displaystyle O}{\overset{\|}{C}}-NH-\underset{\displaystyle R}{\underset{|}{CH}}-\overset{\displaystyle O}{\overset{\|}{C}}-OH + H_2O$$

25. The primary protein structure simply refers to the sequence of amino acids in the protein chain. The primary protein structure is maintained by the peptide bonds formed between the amino acids.

27. The tertiary protein structure refers to the large scale bends and folds in the protein structure. The tertiary protein structure results from interactions between individual amino acid R units that are separated from each other by a large distance along the linear sequence of amino acids.

29. The α-helix structure occurs when the amino acid chain is wrapped into a tight coil, much the same as a spring, with the R groups extending outward. The β-pleated sheet structure has an extended chain that is in a zigzag pattern.

31. Nucleic acids contain a chemical code that specifies the correct amino acid sequences for the creation of proteins.

33. The four bases found within DNA are adenine (A), cytosine (C), guanine (G), and thymine (T).

35. The genetic code links a specific codon to an amino acid.

37. A gene is a sequence of codons within a DNA molecule that codes for a single protein. Genes vary in length from fifty to thousands of amino acids.

39. A chromosome is the structure within a cell that contains the genes.

41. No, cells only produce those proteins that are critical to the cell type and function.

43. a) thymine (T)
    b) adenine (A)
    c) guanine (G)
    d) cytosine (C)

Problems

Carbohydrates

45. a) carbohydrate, monosaccharide
    b) not a carbohydrate
    c) not a carbohydrate
    d) carbohydrate, disaccharide

47. a) hexose
    b) tetrose
    c) pentose
    d) tetrose

49. The linear and ring structures for glucose are:

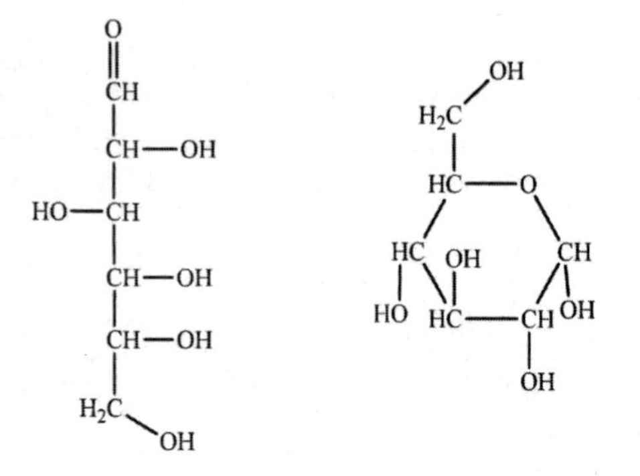

51. The structure of sucrose:

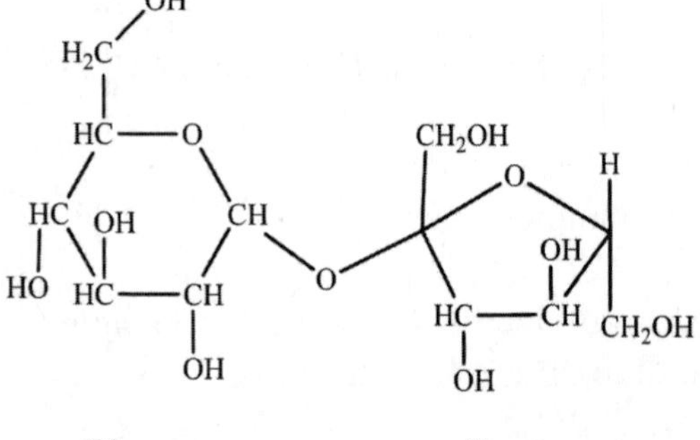

Lipids

53. a) lipid-fatty acid-saturated
    b) lipid-steroid
    c) lipid-triglyceride-unsaturated
    d) not a lipid

55. The block diagram of a triglyceride is:

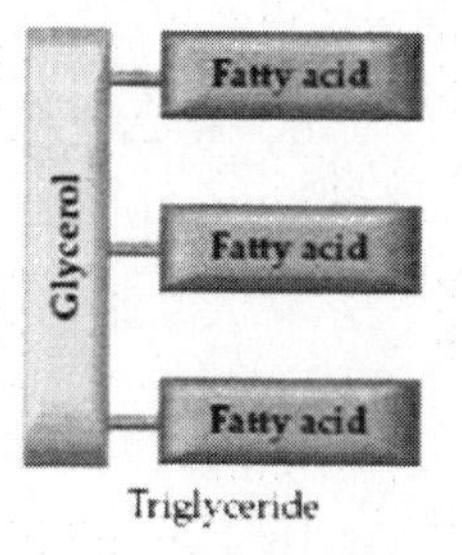

Triglyceride

57. The structure of the triglyceride is below. Because it is a saturated fat, you would expect it to be a solid.

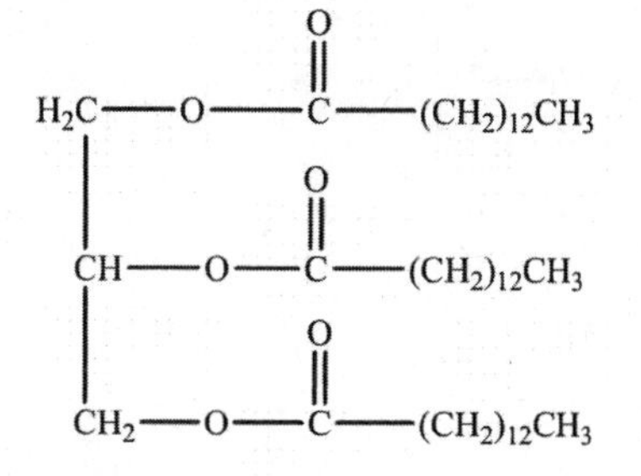

Amino Acids and Proteins

59. a) not an amino acid
    b) amino acid
    c) not an amino acid
    d) amino acid

61. 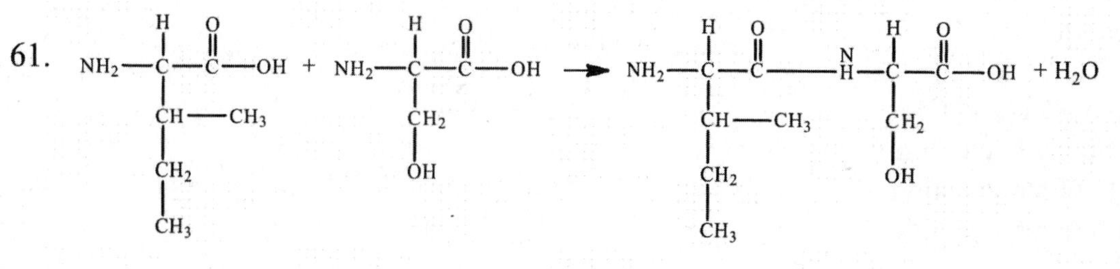

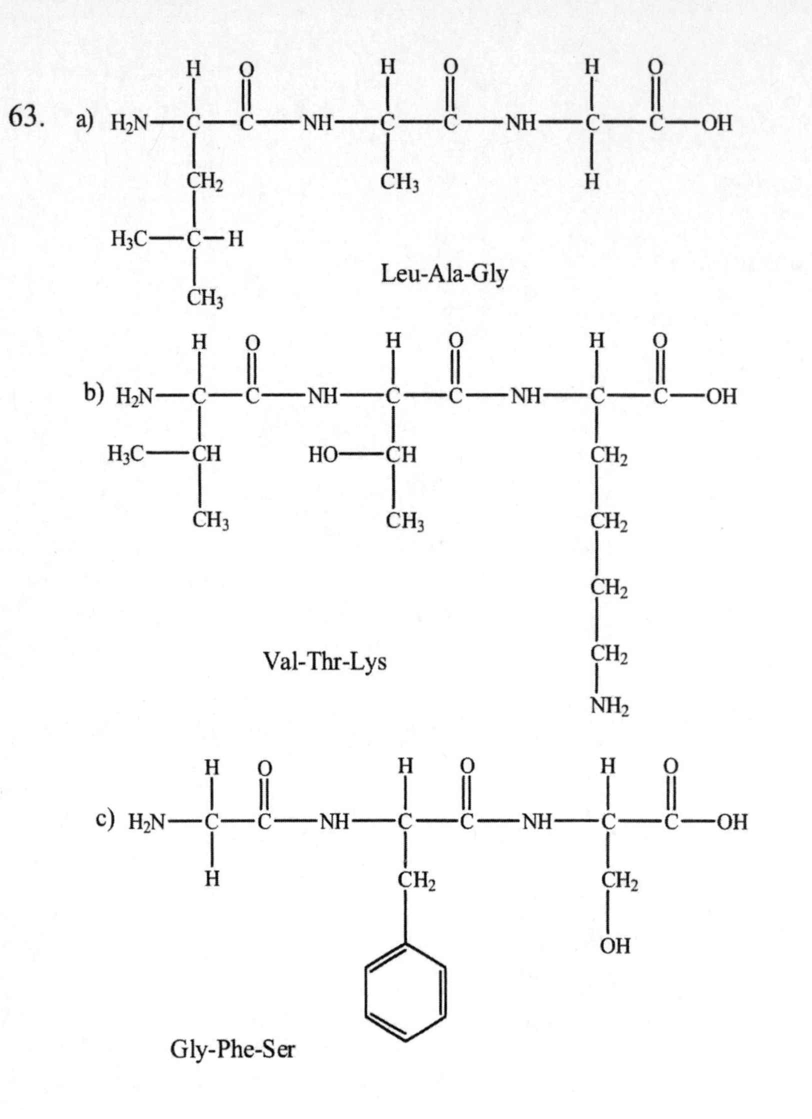

65. This interaction is an example of a tertiary structure because the amino acids involved are a long distance apart in terms of their placement in the chain.

67. A listing of the amino acids in order of their appearance in the chain is a primary structure.

Nucleic Acids

69. a) nucleotide, G
    b) not a nucleotide
    c) not a nucleotide
    d) not a nucleotide

71. The complementary strand of DNA is: T T A C G C G

73. DNA replicates as follows:

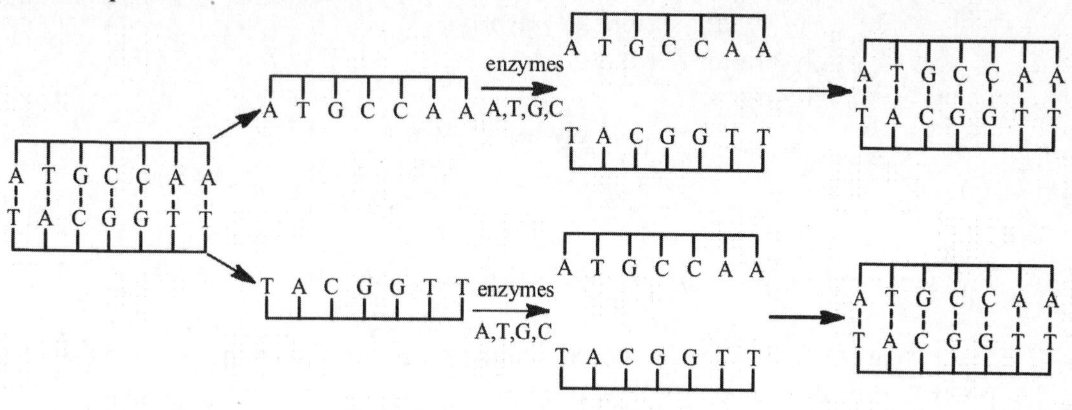

1. Hydrogen bonds break 2. Complementary bases match to each strand 3. Hydrogen bonds reform

Cumulative Problems

75. a) glycoside linkage—carbohydrates
 b) peptide bonds—proteins
 c) ester linkage—triglycerides

77. a) glucose—short-term energy storage
 b) DNA—blueprint for proteins
 c) phopholipids—compose cell membranes
 d) triglycerides—long-term energy storage

79. a) codon—codes for a single amino acid
 b) gene—codes for a single protein
 c) genome—all of the genetic material of an organism
 d) chromosome—structure that contains genes

81. Nitrogen (1): Trigonal Pyramidal
 Carbon (2): Tetrahedral
 Carbon (3): Trigonal Planar
 Oxygen (4): Bent

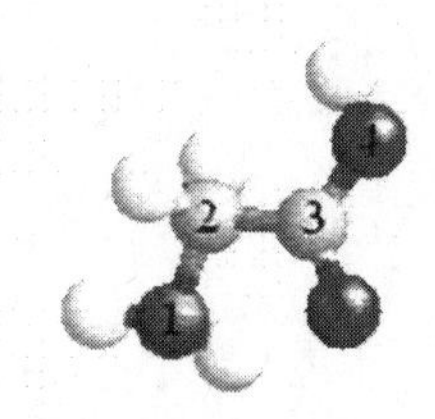

83.

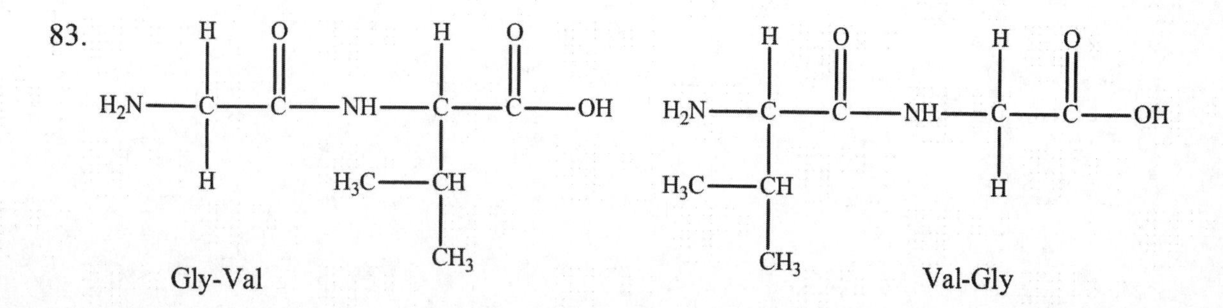

85. Lining up fragment pieces that overlap patterns:

ala-ser-phe-gly-asn-lys
gly-arg-ala-ser-phe
gly-arg
gly-asn-lys-trp
trp-glu-val
glu-val

Protein: gly-arg-ala-ser-phe-gly-asn-lys-trp-glu-val

87. Since each amino acid in the protein requires a codon consisting of three base pairs for synthesis, the number of base pairs is:

51 amino acids x (3 bases/codon) = 153 bases

89. $3.66 \text{ torr} \times \frac{1 \text{ atm}}{760 \text{ torr}} = n\left(0.0821 \frac{\text{L} \bullet \text{atm}}{\text{mol} \bullet \text{K}}\right) 298 \text{ K} \Rightarrow$

$n = 1.97 \times 10^{-4}$

$n = \frac{(\text{mass/molar mass})}{\text{L}} \Rightarrow 1.97 \times 10^{-4} = \frac{0.02388/\text{Molar Mass}}{0.0200} \Rightarrow$

$\text{Molar Mass} = \frac{0.02388}{(1.97 \times 10^{-4})0.0200} = 6.06 \times 10^{3} \text{ g/mol}$

Highlight Problems

91. The actual thymine-containing nucleotide uses the -OH end to bond and replicate; however, with the fake nucleotide having a nitrogen-based end instead, the possibility of replication is halted.